Essential Electronics Series

Introduction to Analogue Electronics

Bryan Hart

Formerly Senior Lecturer in Electronics
University of London

Newnes

OXFORD AMSTERDAM BOSTON LONDON NEW YORK PARIS
SAN DIEGO SAN FRANCISCO SINGAPORE SYDNEY TOKYO

To the memory of my parents
Agnes L. Hart (1892–1976) and Charles H. Hart (1896–1993)
who were born at the dawn of the Age of Electronics

Newnes
An imprint of Elsevier Science
Linacre House, Jordan Hill, Oxford OX2 8DP
225 Wildwood Avenue, Woburn MA 01801-2041

First published 1997
Transferred to digital printing 2002

British Library Cataloguing in Publication Data
A catalogue record for this book is available from the British Library

ISBN 0 340 65248 9

For information on all Newnes publications
visit our website at www.newnespress.com

Printed and bound in Great Britain by Antony Rowe Ltd, Eastbourne

Contents

by 'doping' the crystal. This means introducing into the crystal, normally via a diffusion process, a small (e.g. 1 part in 10^7) controlled quantity of an impurity atom.

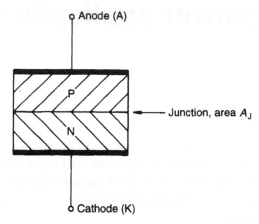

Fig. 1.1   Schematic cross-section of a PN junction diode. The darkened end-regions are metallic contacts

P-type material is formed by doping a crystal with an 'acceptor' impurity, which is one that supplies holes. Thus, in a P region most of the charge carriers available for conduction are positive; holes are 'majority carriers'. There are still some negative charge carriers (electrons) available, through the thermal disruption of bonds, but these 'minority carriers' are sparsely distributed. N-type material is formed by doping a crystal with a 'donor' impurity, which is one that donates electrons. For an N region the majority carriers are negative (electrons) and holes are the minority carriers.

In the design of a PN junction diode the concentration of holes in the P region is made much greater than the concentration of electrons in the N region. What happens in the junction region when an external voltage is applied to the device is well-documented in textbooks on solid-state device electronics (*see,* for example, Till and Luxon (1982)) so only a simplified explanation is given here.

If a voltage is applied to the PN junction diode to make the P region positive with respect to the N region the diode is said to be 'forward-biased'. Holes are encouraged to flow from the P region where they are plentiful, across the junction and into the N region where they are in short supply. Similarly, electrons flow from the N region, where they are plentiful, to the P region where they are sparse in number. Because of the gross asymmetry in doping levels, the terminal current is mainly due to the flow of holes from P to N. The energies of the carriers are distributed in approximately the same way as the energies of the molecules of an ideal gas so the current tends to increase exponentially with applied voltage.

Under 'reverse bias' conditions, the polarity of the applied voltage (P region made negative with respect to the N) encourages holes to flow across the junction from N to P and electrons from P to N. Thus, carriers are sought from regions of short supply and the reverse current is very small, being limited by the rate at which bonds are broken by thermal energy.

The junction diode characteristic

Figure 1.2 shows a general diode symbol. I indicates the direction of conventional current flow when an external voltage, V_{AK} is applied. Thus I corresponds to the

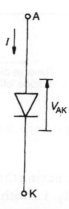

Fig. 1.2　General symbol for diode, without restriction on its characteristic. (The triangular part of the symbol is sometimes shaded-in.) The lettering, not part of the symbol, shows convention for positive I, V

flow of positive charge into the anode. The equation that applies to an 'ideal logarithmic diode' is,

$$I = I_S(e^{V_D/V_T} - 1) \tag{1.1}$$

In this, which is the simplest mathematical expression that is a good fit to most practical diodes, the symbols have the following meanings:

I_S = diode reverse saturation current = $A_J J_S$, where A_J = junction area, J_S = junction current density. J_S is dependent on material type, thickness doping of P and N regions and temperature.

e = base of Naperian logarithms ($= 2.7182 \ldots$),

V_{AK}, V_D = diode (voltage) drop,

V_T = 'thermal voltage' = k(T/q), where k = Boltzmann's constant = 1.38×10^{-23} J/K, T = absolute temperature in Kelvin (°C + 273), q = magnitude of electronic charge = 1.602×10^{-19} C.

The voltage unit V_T occurs repeatedly in the characterization of semiconductor devices and is approximately 25 mV at $T = 293$ K. Unless otherwise stated, that value will be used from now on because it is easy to remember and convenient for rough calculations.

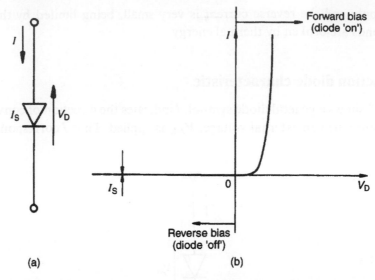

(a)   (b)

Fig. 1.3   (a) Symbol for ideal logarithmic diode. (b) Static terminal characteristic (*I–V* curve) for (a)

The symbol reserved for a diode having the characteristic specified by Equation (1.1) is shown in Fig. 1.3(a): it is Fig. 1.2 with the addition of the parameter I_S and the replacement of V_{AK} by V_D.

A sketch of Equation (1.1) is shown in Fig. 1.3(b). On it, I_S appears to lie on the axis $I = 0$ because it is so small on the usual scale chosen to display I. To emphasize the existence of a finite, though small, I_S, Fig. 1.3(b) is frequently shown with an expanded scale for $I < 0$, as indicated in Fig. 1.4. Remember though, the curve is continuous, i.e. there is no abrupt change in shape as it passes through the origin.

In forward-bias and with $V_D > 4V_T$ (i.e. 100 mV), or equivalently $I \gg I_S$, Equation (1.1) can be rewritten as,

$$I = I_S e^{V_D/V_T} \tag{1.2}$$

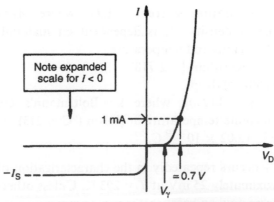

Fig. 1.4   Redrawn version of Fig. 1.3(b) with expanded reverse scale. V_γ indicates a forward voltage of special interest

Unlike the gentle curve of e^x in a mathematics textbook, the I–V characteristic of a practical diode shows a sharp 'corner' when drawn on a scale suitable for most practical situations.

With low power Si diodes $I_S \sim 1\text{pA}$ (i.e. 10^{-12} A) and I is negligible in comparison with usual operating current levels until V_D exceeds a certain 'threshold' or 'cut-in' voltage, denoted by V_γ in Fig. 1.4. Usually we assume $V_\gamma = 0.6$ V. A useful rule-of-thumb, for $V_D > V_\gamma$, is $I = 1$ mA when $V_D = 0.7$ V. For low power Ge diodes: $I_S \sim 1\ \mu\text{A}$; $V_\gamma \sim 0.2$ V; $I \simeq 1$ mA when $V_D = 0.3$ V. Equation (1.2) is also valid for instantaneous values of time-varying voltage, v_D and current, i. Thus, a general form for Equation (1.2) is,

$$i = I_S e^{v_D/V_T} \tag{1.3}$$

Example 1.1 _____

Calculate the change in voltage drop when the forward current ($\gg I_S$) in a diode changes by a factor of 2 (an 'octave change').

Solution

Using Equation (1.2) twice,

$$I_1 = I_S e^{V_{D1}/V_T}$$

$$I_2 = I_S e^{V_{D2}/V_T}$$

$$\therefore (I_2/I_1) = e^{(V_{D2}-V_{D1})/V_T} = e^{\Delta V_D/V_T}$$

Taking natural logarithms of each side,

$$\Delta V_D = V_T \log_e(I_2/I_1)$$

If $I_2 = 2I$, $\Delta V_D = V_T \log_e 2 = +17.3$ mV
If, however, $I_2 = 0.5I_1$, then $\Delta V_D = -17.3$ mV

For reverse bias and $V_D < -4V_T$,

$$I = -I_S \tag{1.4}$$

The negative sign occurs because I is taken as positive when directed from A to K, as indicated by the full-line arrow in Fig. 1.5(a). The actual direction of current flow is shown dotted. An equivalent d.c. circuit for the diode under these conditions is shown in Fig. 1.5(b).

Equation (1.4) is not necessarily a good description of device operation for time-varying applied voltages, for the reason given in Section 1.3.

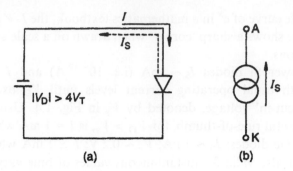

Fig. 1.5   (a) Diode with reverse bias: dotted line indicates direction of *actual* current flow. (b) Simple equivalent circuit for diode in (a)

Temperature dependence of *I–V* curve

Figure 1.6 indicates the general effect of temperature (T) change on the *I–V* curve. As a memory-aid, recall that the diode conducts more in both directions when hotter. For a constant forward current,

$$(dV_D/dT) = -a \qquad (1.5)$$

Normally, 'a' is taken as constant at about 2 mV/°C. Thus, in Fig. 1.6,

$$|\Delta V_D| \simeq 2(T_2 - T_1)\,\text{mV} \qquad (1.6)$$

This relationship has been exploited in the design of rudimentary solid-state thermometers.

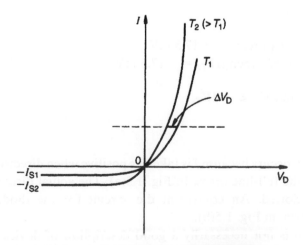

Fig. 1.6   Showing the temperature dependence of the *I–V* curve

For I_S,

$$(1/I_S)/(dI_S/dT) = b \qquad (1.7)$$

With Si diodes, $b \simeq 0.7$.

Integrating Equation (1.7) gives,

$$I_S(T_2) = I_S(T_1)\,e^{b(T_2 - T_1)} \qquad (1.8)$$

This discussion introduces the way device data sheet specifications quantify the temperature dependence of a parameter X. The terms 'Temperature Coefficient of X', 'Tempco X' and the letters TCX are used. However, two meanings have been given to these. The first is the absolute variation of X with T: in this case TCX = dX/dT. The second is the fractional variation of X with T: here, TCX = $(1/X)(dX/dT)$, which is usually expressed as a percentage per °C or as parts per million per °C (ppm/°C). (Note: 100 ppm = 0.01%). If the precise meaning of TCX is not stated it can be deduced from the units used to express it.

Diode models

Equation (1.1), or an *I–V* curve such as Fig. 1.3(b), must be used for finding the forward voltage drop across a diode if it is connected in a circuit in which the supply voltage is comparable with that drop. However, if the applied voltage greatly exceeds the drop, as in rectifier circuits, the arithmetic complication of dealing with an exponential function can be avoided by assuming simpler characteristics for the diode. These are 'piecewise-linear' in nature. Curved characteristics are approximated by linear sections.

The simplest model for a diode is the 'ideal' piecewise-linear model, represented by the *I–V* graph of Fig. 1.7(a). For $V_D > 0$, the diode is a short circuit (a closed switch): for $V < 0$, the diode is an open circuit (an open switch).

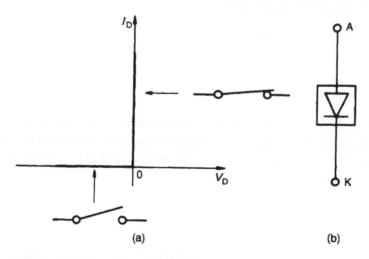

(a)   (b)

Fig. 1.7   (a) Ideal piecewise-linear characteristic. (b) Circuit model

A general diode symbol enclosed by a square box (Fig. 1.7(b)) is convenient for symbolizing this characteristic.

A less crude approximation is the assumption that the diode drop is constant at a value δ (e.g. 0.7 V) when the diode is on and the current is zero when it is off ($V_D < 0.7$ V). Figure 1.8 shows the associated characteristic and a circuit representation. This model is useful in some types of logic circuit. An incremental model of the diode characterizes its behaviour for small changes about a 'quiescent', or 'operating', point Q.

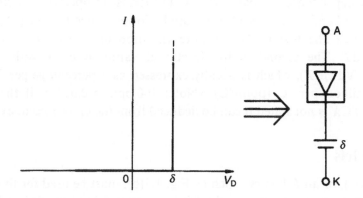

Fig. 1.8   Improved piecewise-linear model

Suppose a change in applied voltage causes the quiescent point to move from V_Q, I_Q to $V_Q + v_q, I_Q + i_q$

Following the procedure of Example 1.1,

$$1 + (i_q/I_Q) = e^{v_q/V_T} \tag{1.9}$$

Using the exponential series but retaining only the three largest terms, on the assumption that (v_q/V_T) is small gives,

$$1 + (i_q/I_Q) = 1 + (v_q/V_T) + (v_q^2/2V_T^2) \tag{1.10}$$

or,

$$i_q = (v_q I_Q/V_T)[1 + (v_q/2V_T)] \tag{1.11}$$

The second term in the square brackets can be ignored if $(v_q/2V_T) \ll 1$

Now, in engineering calculations the condition $a \ll b$ is considered to be met if $a \leqslant (b/10)$. Applying that criterion here, the requirement is $v_q \leqslant 5$ mV. Equation (1.11) then becomes,

$$i_q = v_q I_Q/V_T = v_q/r_d \tag{1.12}$$

where,

$$r_d = V_T/I_Q \tag{1.13}$$

r_d is the 'incremental' or 'slope' resistance and has the same ohmic value for all ideal logarithmic diodes operating at a specified I_Q and T.

The description 'diode resistance', without qualification, is ambiguous. It could mean V_Q/I_Q, but that has a different value for each I_Q and it certainly does not relate i_q to v_q in a useful way.

Figure 1.9(a) illustrates the mathematical argument. If $v_q \leqslant 5$ mV, this condition being independent of V_Q, then i_q is proportional to v_q, the constant of proportionality being the inverse of the slope resistance r_d ($= V_T/I_Q$). The expression for r_d can be obtained directly by differentiating Equation (1.2) (*see* Problem 5) but that procedure does not give a precise meaning to 'small change'. Figure 1.9(b) shows an incremental model of the diode.

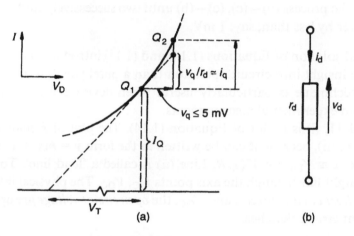

Fig. 1.9   (a) Characterization of slope resistance r_d. (b) Incremental model of diode

Load lines

To determine the current I ($\gg I_S$) for the circuit of Fig. 1.10, two simultaneous equations must be satisfied. These describe the *I–V* requirements of each of the components, D and R.

For D,

$$I = I_S e^{V_D/V_T} \tag{1.14}$$

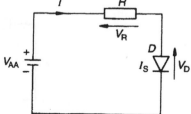

Fig. 1.10   Basic series circuit with *D* forward-biased

For R,

$$I = V_R/R = (V_{AA} - V_D)/R = -(V_D/R) + (V_{AA}/R) \qquad (1.15)$$

Equating Equations (1.14) and (1.15) to find the value $I = I_Q$ that satisfies both, leads to an equation that cannot be solved directly. However, it can be solved by an 'iterative' process using a pocket calculator. The procedure is as follows:

(a) assume an initial value V_{Q1} for V_D (say, 0.6 V).
     Use Equation (1.15) to calculate $I = I_{Q1}$;
(b) substitute I_{Q1} into Equation (1.14) to find a new value V_{Q2};
(c) substitute V_{Q2} into Equation (1.15) to find I_{Q2};
(d) repeat the process (b)→(c), (c)→(b) until two successively calculated values of V_Q differ by less than, say, 1 mV.

A graphical solution of Equations (1.14) and (1.15) introduces the 'load line' and gives more insight into circuit operation than a purely analytical approach. The load line technique is particularly useful when device characteristics are only known from experimental measurements.

In Fig. 1.11, (i) is a plot of Equation (1.14). The plot of Equation (1.15) is a straight line, (ii), because it can be written in the form $y = mx + c$, where $y \equiv I$, $m \equiv -1/R$, $x \equiv V_D$, $c \equiv V_{AA}/R$. Line (ii) is called a 'load line'. To plot it, just draw a straight line through the axis points I_{SC}, V_{OC}. The physical interpretations of I_{SC} the *short circuit current*, and V_{OC}, the *open circuit voltage* are apparent from the adjacent circuit sketches.

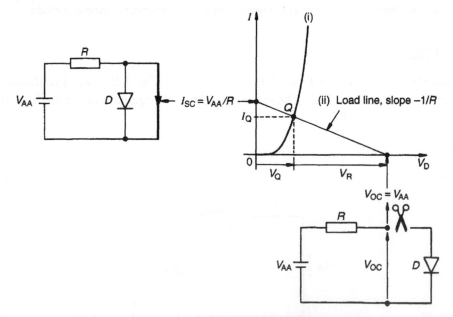

Fig. 1.11   'Load-line' construction for finding quiescent current I_Q. Adjacent circuits define I_{SC}, V_{OC}

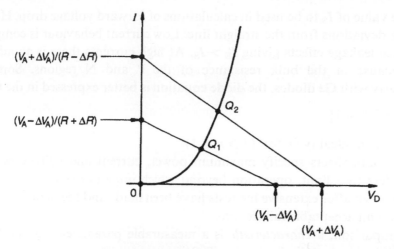

Fig. 1.12   Load lines illustrating limit conditions

The point of intersection, Q, gives the operating voltage, V_Q, and current I_Q. If $V_{AA} \gg V_Q$, then $I_Q \simeq I_{SC}$ and the use of the diode model of Fig. 1.7 is justified.

The value of the load line technique in indicating operating limits is demonstrated in Fig. 1.12.

If, in Fig. 1.10, R and V_A are subject to variations $\pm\Delta R$, $\pm\Delta V_A$, respectively, then all possible load lines must lie between the two shown. Operating points Q_1, Q_2 correspond, respectively, to minimum and maximum currents.

Practical points

Up to this point, only the ideal logarithmic diode has been considered. Some features of practical diodes are now discussed.

For a practical diode, there is normally a straight-line relationship, indicated by the bold line in Fig. 1.13, between I plotted to a logarithmic scale and V_D, plotted to a linear scale. This is a semilog plot. The extrapolated intercept on the I axis

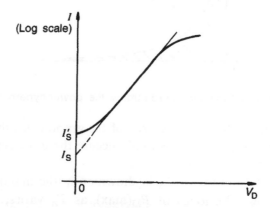

Fig. 1.13   Practical diode characteristic plotted to log/linear scales

gives the value of I_S to be used in calculations of forward voltage drop. However, there are deviations from the straight line. Low current behaviour is complicated by surface leakage effects giving $I'_S > I_S$. At high currents there is an added V_D drop because of the bulk resistance of the P and N regions. Sometimes, particularly with Ge diodes, the diode equation is better expressed in the form,

$$I = I_S[e^{V_D/nV_T} - 1] \tag{1.16}$$

In this, n is an 'ideality factor' $(2 \geqslant n \geqslant 1)$.

Device data sheets specify maximum power, current and voltage *ratings*. A rating refers to a limit, operation beyond which impairs the serviceability of a device. It is set after extensive life tests have been made and the manufacturer has decided on an acceptable failure rate.

By comparison, a *characteristic* is a measurable parameter, e.g. the forward voltage drop across a diode at a specified current.

The power, P_D, dissipated by a diode is given by $P_D = I_Q V_Q$. The data sheet gives a maximum permitted power, $P_D(\text{max})$, that is dependent on the construction, packaging and environment of the device. If $P_D(\text{max})$ is exceeded for a significant time interval the diode might be irreparably damaged. To ensure this does not happen a design rule is,

$$I_Q V_Q < P_D(\text{max}) \tag{1.17}$$

The plot of $P_D(\text{max})$ on an *I–V* graph is a hyperbola (Fig. 1.14) which is easily drawn. Thus, if $P_D(\text{max}) = 250$ mW, one point on the curve is $I_Q = 500$ mA, $V_Q = 0.5$ V.

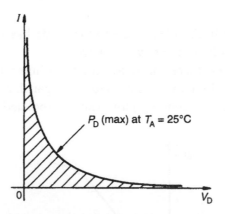

Fig. 1.14   Operation must be in shaded area below the 'power hyperbola' $IV_D = P_D(\text{max})$

Operation must be confined to the area of the graph beneath the appropriate hyperbola. The word 'appropriate' is used because there is a separate hyperbola for each value of ambient temperature, T_A.

The variation of $P_D(\text{max})$ with T_A is shown on a 'thermal derating' diagram (Fig. 1.15). This shows the locus of $P_D(\text{max})$, as T_A varies, when the device junction temperature, T_J, is at its maximum permitted value, T_{JM}. T_{JM} is set by

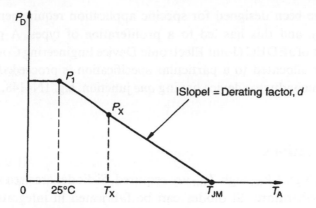

Fig. 1.15   Power derating curve. T_A = ambient temp., T_{JM} = max. junction temp., both in °C

the semiconductor material and the packaging, and is 85°C for Ge devices and 175°C for industrial-grade Si devices.

For $T_A = T_X$ (>25°C) the maximum power dissipation, P_X, is related to that (P_1) at $T_A = 25°C$ by the expression,

$$P_X = P_1 - d(T_X - 25) \tag{1.18}$$

The parameter 'd' (mW/°C, W/°C) is the 'thermal derating factor' and is given by

$$d = P_1/(T_{JM} - 25) \tag{1.19}$$

The inverse of d is the 'thermal resistance' θ_{JA}, which is expressed in °C/mW, °C/W.

The power management of semiconductor devices is considered in greater depth in a companion volume.

The reverse current in a practical diode is not constant for all values of reverse voltage that exceed $4V_T$. When the voltage reaches a critical value a junction breakdown mechanism occurs and a large current flows (Fig. 1.16). The breakdown voltage, denoted by BV_R, is specified as that voltage at which the reverse current is equal to a standard test value. With 'signal diodes' and diodes intended for use in rectifier circuits BV_R must not be exceeded. However, some special diodes, designed to have a low BV_R, are operated in the breakdown region as discussed in Section 1.2.

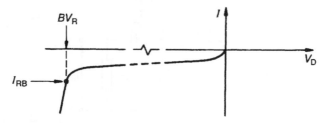

Fig. 1.16   Diode breakdown characteristic

Diodes have been designed for specific application requirements, e.g. low I_S, high $P_D(\text{max})$, and this has led to a proliferation of types. A popular coding method is that of JEDEC (Joint Electronic Device Engineering Council). In this, a diode number allocated to a particular specification is proceeded by 1N, which refers to a semiconductor device having *one* junction, e.g. 1N4148, a popular low-cost Si diode.

Diode applications

Where low I_S, high BV_R, high T_{JM} are required, Si diodes are used in preference to Ge types. Furthermore, Si diodes can be fabricated in integrated circuits. Ge diodes are preferred in some rectifying circuits when a low forward voltage drop is important. In the small sample of applications that follows, it is assumed that Si diodes are employed.

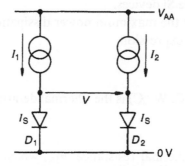

Fig. 1.17   Monolithic diode-pair working as a thermometer

Figure 1.17 shows the basic circuit of an accurate thermometer for human and industrial use. D_1, D_2, part of a monolithic integrated circuit, may be assumed to have identical characteristics. However, they operate at different constant currents. It can be shown that $V \propto T$ (*see* Problem 8).

A simple electronic attenuator for gain-control applications is shown in Fig. 1.18. R_S and the slope resistance, r_d, of D form a potentiometer network that

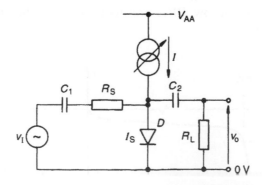

Fig. 1.18   Simple diode attenuator

attenuates a sinusoidal input v_1 of peak value V_{im} and angular frequency ω. Design requirements are: $(1/\omega C_1) \ll R_S$; $(1/\omega C_2) \ll R_L$; $R_L \gg r_d$. It follows that,

$$v_o = [V_{im} \sin \omega t] \times [r_d/(r_d + R_S)] \tag{1.20}$$

A desired attentuation is achieved by choice of I, which controls r_d.

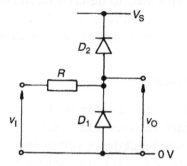

Fig. 1.19   Diode limiter

In considering Figs 1.19 and 1.20, use is made of the diode model of Fig. 1.8. Figure 1.19 shows a protection circuit similar to that used in CMOS logic gates. D_1, D_2 ensure that $(V_S + \delta) \geqslant v_o \geqslant -\delta$.

A simple two-input AND gate (positive logic convention), together with illustrative circuit waveforms, is shown in Fig. 1.20.

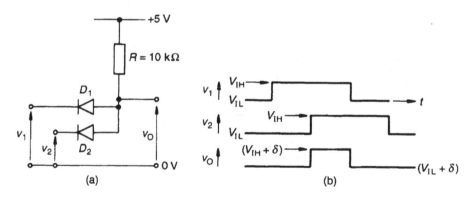

Fig.1.20   (a) Logic (AND) gate. (b) Waveforms for (a)

If $v_1 = V_{IH}$, $v_2 = V_{IL}$, then D_2 is on and D_1 is off, with $V_D = -[V_{IH} - (V_{IL} + \delta)]$. The roles of D_1, D_2 are reversed if $v_1 = V_{IL}$, $v_2 = V_{IH}$; $v_0 = (V_{IH} + \delta)$ only if $v_1 = v_2 = V_{IH}$.

1.2   THE ZENER DIODE

Two well-known physical processes are responsible for reverse voltage breakdown in a PN junction diode, the 'Zener effect' and the 'Avalanche effect'. With

$P_D(max) = 500$ mW at $T_A = 25°C$, derating to zero at $T_J = 175°C$.
Calculate $P_D(max)$, $I_Z(max)$ at $T_A = 50°C$.

Solution

Equations (1.18) and (1.19) apply if we write: $P_X = P_D(max)$ at $50°C$;
$P_1 = P_D(max)$ at $25°C = 500$ mW; $T_X = 50°C$; $T_J = T_{JM} = 175°C$.
Using Equation (1.19), $d = (500/150)$ mW/°C. Substituting this value into Equation (1.18) gives $P_X = 417$ mW.
Now $P_D = I_Z V_Z$, so substituting for V_Z from Equation (1.21a), inserting numerical data and operating in mW,

$$P_D = 0.02 I_Z^2 + 5 I_Z$$

The following condition must be met,

$$0.02 I_Z^2 + 5 I_Z \leqslant 417$$

$I_Z(max)$ is found by taking the equality sign and solving the resultant quadratic equation,

$$I_Z(max) \leqslant 66 \text{ mA}$$

Zener applications

A popular application of zener diodes and one which is considered in more detail, below, is the simple shunt stabilizer shown in Fig. 1.24. V_O stays nearly constant if I_L increases because the required current is diverted from the zener with only a small change in zener voltage.

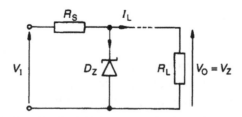

Fig. 1.24   Basic shunt stabilizer (or 'regulator')

Figure 1.24 can also be regarded as a limiter if V_I is replaced by a large amplitude time-varying signal. Figure 1.25 shows a direct-coupling circuit that offers voltage level shift without significant attentuation (Problem 11). A sensitive voltmeter (M) can be protected from potentially destructive overloads if an appropriately chosen zener diode, D_Z is connected across it, as shown in Fig. 1.26.

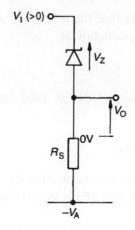

$V_I \ (>0)$

V_Z

V_O

0V

R_S

$-V_A$

Fig. 1.25 Zener diode coupling circuit

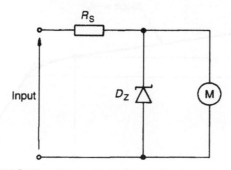

R_S

Input

D_Z

M

Fig. 1.26 Meter protection circuit

Returning to the shunt stabilizer of Fig. 1.24, an equivalent circuit, which incorporates the diode model of Fig. 1.23, is shown in Fig. 1.27. At node X,

$$I_R = I_Z + I_L \tag{1.22}$$

Substituting for I_R, I_Z, gives,

$$(V_I - V_Z)/R_S = [(V_Z - V_{ZS})/r_z] + I_{ZS} + I_L \tag{1.23}$$

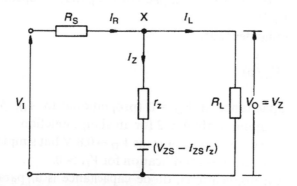

R_S I_R X I_L

I_Z

V_I r_z R_L $V_O = V_Z$

$(V_{ZS} - I_{ZS} r_z)$

Fig. 1.27 Equivalent circuit of shunt stabilizer in Fig. 1.24

From Equation (1.23), it follows that the change ΔV_Z, not necessarily small, that occurs for a change ΔI_L with V_I constant is,

$$\Delta V_Z = -\Delta I_L r_z R_S/(r_z + R_S) \simeq -r_z \Delta I_L \tag{1.24}$$

The approximation in Equation (1.24) is valid because $r_z \ll R_S$ in practice. Furthermore, for I_L constant,

$$(\Delta V_Z/\Delta V_I) = r_z/(R_S + r_z) \simeq r_z/R_S \tag{1.25}$$

Equation 1.25 is also valid for R_L constant, since $R_L \gg r_z$.
    A regulation characteristic (a plot of V_O vs I_L) is shown in Fig. 1.28.

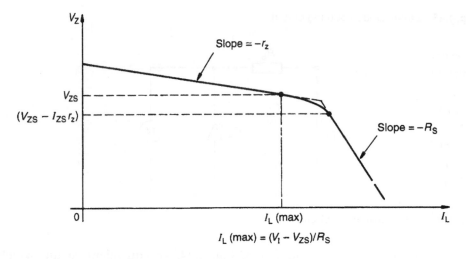

Fig. 1.28   Regulation characteristic for Fig. 1.27

1.3 VARICAP DIODES

A dipole layer of charge straddles a PN junction (Till and Luxon, 1982) and gives rise to a 'depletion' or 'transition' capacitance, C_j, that is dependent on operating voltage (*see* Fig. 1.29).
    Simple theory indicates that,

$$C_j = C_j(0)/\sqrt[n]{1 - (V_D/\phi)} \tag{1.26}$$

In this: $C_j(0)$ = capacitance at $V_D = 0$ (proportional to A_J); ϕ = a constant $\simeq$ 0.8 V for Si; $n = f$(doping profile) $\simeq 2$ for an abrupt junction.
    Equation (1.26) predicts that $C_j \to \infty$ as $V_D \to 0.8$ V but simple theory, which is adequate for $V_D < 0$, requires modification for $V_D > 0$.
    In many practical applications, diode capacitance is a 'parasitic' element in circuit operation. Current flowing in it when a diode is subjected to time-varying

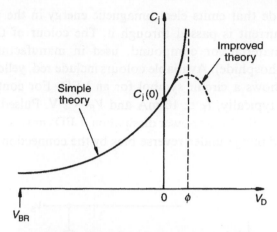

Fig. 1.29 Capacitance–voltage characteristic of a junction diode

reverse voltages can far exceed I_S, so the simple model of Fig. 1.5(b) is no longer valid for time-varying signals.

However, the 'varicap' (variable capacitance) diode, or 'silicon tuning diode', is purposely designed to exploit the variation of C_j with reverse voltage. Its symbol is shown in Fig. 1.30(a) and it is always used in reverse bias. A typical application, in radio receivers, is shown in Fig. 1.30(b). C_D is a decoupling capacitor (e.g. 10 nF) the reactance of which is insignificant compared with that of R_S or C_j at the frequency used. The resonant frequency, f_R, of the parallel tuned circuit is,

$$f_R = 1/2\pi\sqrt{L(C + C_j)} \tag{1.27}$$

Adjustment of V_{AA} gives rise to variation of C_j, and hence f_R.

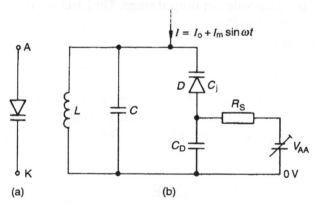

Fig. 1.30 (a) Varicap diode symbol. (b) Varicap tuning application

1.4 THE LIGHT EMITTING DIODE (LED)

The LED is a semiconductor counterpart of a low voltage, low power, electric lamp and is widely used for visual read-out in electronic instruments. The device is a PN junction diode that emits electromagnetic energy in the visible spectrum when a forward current is passed through it. The colour of the light emitted depends on the material, or compound, used in manufacture (e.g. gallium arsenide, gallium phosphide). Available colours include red, yellow, green, blue.

Figure 1.31(a) shows a circuit symbol for an LED. For continuous forward current operation, typically, $I_F \simeq 10$ mA and $V_F = 2$ V. Pulsed operation gives increased light output for lower power dissipation. LEDs have a low BV_R and can be protected from damage under reverse bias by the connection of a Si diode in inverse parallel.

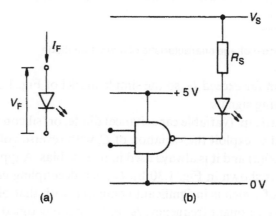

(a)   (b)

Fig. 1.31   (a) Symbol for an LED. (b) Logic state indicator using an LED

Figure 1.31(b) shows the use of an LED as a logic-state indicator, driven by a TTL gate with an open-collector output stage. The LED is only on if the output of the gate is at its low value V_{OL}.

The design equation is,

$$R_S = (V_S - V_F - V_{OL})/I_F \qquad (1.28)$$

The tolerance on R_S is non-critical, $\pm 10\%$ being acceptable.

Combining a number of LEDs into a single package gives rise to the following types of visual display: dot-matrix, star-burst, light-bar, seven-segment.

1.5   THE SCHOTTKY DIODE

The metal used to connect the external leads of a PN junction forms contacts with the semiconductor material that are designed to be 'ohmic' or 'non-rectifying'. Thus, they present the same low resistance to current flow from metal to semiconductor as vice versa. However, by an appropriate choice of metal, semiconductor and fabrication method, a metal–semiconductor contact can be made that has the exponential characteristic of a PN junction diode. A device that

Fig. 1.32 Symbol for a Schottky diode

exploits this property is called a 'Schottky' or 'hot-carrier' diode and the symbol used for it is shown in Fig. 1.32.

Compared with a low power Si diode of similar physical dimensions the Schottky has: lower V_γ ($\simeq 0.3$ V) and a lower forward voltage drop at a given forward current; higher I_S; lower BV_R; and capability of much faster switching performance because of the different charge transport mechanism involved.

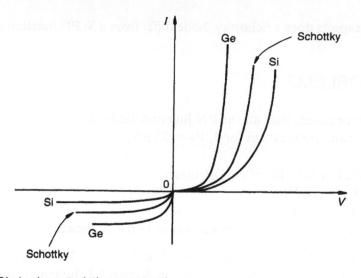

Fig. 1.33 Diode characteristics compared

Figure 1.33 compares the $I–V$ curves of Si, Ge and Schottky diodes. Discrete Schottky diodes are used in high-speed waveform generation and shaping circuits. Furthermore, their ease of fabrication, in IC form, leads to their application in the design of fast IC logic gates.

1.6   SELF ASSESSMENT TEST

1 What is the principal electrical characteristic of a PN junction diode?

2 In what way does the temperature affect the I–V curve of a diode?

3 What is meant by the 'incremental resistance' of a diode?

4 Through what axis points does a load line pass?

5 What is a 'power hyperbola'?

6 In what units is 'thermal resistance' measured?

7 Why is the reverse current rating of a zener diode less than the forward current rating?

8 Does the capacitance of a Varicap diode increase or decrease if the reverse voltage changes from -2 V to -3 V?

9 In what respects does a Schottky diode differ from a Si PN junction diode?

1.7   PROBLEMS

(Assume, throughout, that all the PN junction diodes have an ideal logarithmic characteristic and that, except for 8, $V_T = 25$ mV).

1 A Si diode has $I_S = 10^{-16}$ A. Calculate:
   (a) I at $V_D = 0.6$ V;
   (b) V_D at $I = 1$ mA;
   (c) ΔV_D if $I\,(\gg I_S)$ changes by a factor of 10 (i.e. a 'decade').

2 In the circuit of Fig. 1.34: $I_{S1} = 10^{-16}$ A; $I_{S2} = 4 \times 10^{-16}$ A. Calculate I_1, using the iterative approach described in Section 1.1 (hint: start by replacing D_1, D_2 by a composite equivalent).

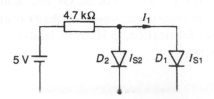

Figure 1.34

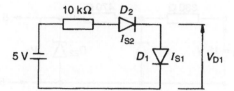

Figure 1.35

3 For Fig. 1.35: $I_{S1} = 4$ pA; $I_{S2} = 2$ pA. Calculate V_{D1}.

4 Equation (1.8) can be written in the alternative form,

$$I_S(T_2) = I_S(T_1) \, 2^{(T_2-T_1)/\theta_X}$$

State, in words, the meaning of this.
    Find θ_X in terms of the parameter b of Equation (1.8). Hence, calculate θ_X
for $b = 0.07$.

5 Establish Equation (1.13) by differentiating Equation (1.2).

6 If D is reversed in Fig. 1.10, show that $V_D = -V_{AA} + I_S R$ using a load line
construction.

7 Referring to Fig. 1.15, show that alternative expressions for P_X are:
    (a) $P_X = d(T_{JM} - T_X) = (T_{JM} - T_X)/\theta_{JA}$
    (b) $P_X = P_1(T_{JM} - T_X)/(T_{JM} - 25)$

8 Referring to Fig. 1.17, assume $I_1 = 100 \ \mu A$, $I_2 = 300 \ \mu A$.
    Show that $(\Delta V/\Delta T) = 94.6 \ \mu V/°C$. (Use the values of k, q given in Section
    1.1).

9 Assume the following for Fig. 1.18: v_I is a 10 kHz sinusoidal signal of peak
    amplitude 25 mV; $R_S = 1$ kΩ; $R_L = 10$ kΩ.
    Choose suitable values for I, C_1, C_2 if the peak value of v_o is 5 mV.

10 The zener diode of Example 1.2 is used at $T_A = 50°C$ in the circuit of Fig. 1.24
    with $V_I = 15$ V. Calculate the maximum permitted value of I_L and an
    appropriate value for R_S, if the minimum zener current is 5 mA.

11 Show that $(\Delta V_o/\Delta V_I) = R_S/(R_S + r_z)$ for the circuit of Fig. 1.25.

12 Derive Equation (1.24) from Equation (1.23).

13 Figure 1.36 shows a two-stage shunt stabilizer. D_{Z1} has $V_{Z1} = 10$ V,
    $r_{z1} = 10 \ \Omega$; D_{Z2} has $V_{Z2} = 5.1$ V, $r_{z2} = 20 \ \Omega$. Calculate ΔV_O for $\Delta V_S =
    \pm 1$ V. (Justify sensible approximations made in the calculation.)

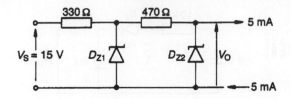

Figure 1.36

14 Varicap data for the circuit of Fig. 1.30 are as follows: $C_j = 25\,\text{pF}$ for $V_{AA} = 0.5\,\text{V}$; $C_j = 425\,\text{pF}$ for $V_{AA} = 15\,\text{V}$.
Calculate $f_R(\text{max})/f_r(\text{min})$ for $C = 25\,\text{pF}$ if V_{AA} is varied over the range 0.5 V to 15 V.

15 Calculate a value of R_S, in the E12 range, that gives $I_F \simeq 10\,\text{mA}$ for the circuit of Fig. 1.31(b) when $V_S = 5\,\text{V}$. Assume $V_F = 2\,\text{V}$, $V_{OL} = 0.2\,\text{V}$.

2 Rectification and power supplies

Direct voltage (d.c.) supplies, such as batteries, are required for electronic devices capable of producing magnified copies (i.e. amplification) of electrical signals that can vary continuously with time, e.g. the output of a microphone. However, batteries need periodic recharging or replacement and are available with only a restricted range of terminal voltages (e.g. 1.5 V, 3 V, etc.). Consequently, batteries are used mainly in low power portable equipment such as personal audio-cassette players. A constant supply of significant amounts of electrical energy is generally only available from an alternating current (a.c.) source, e.g. a.c. mains. The reasons for this are the ease, efficiency and cost of generation, voltage-transformation and transmission.

'Rectification' is the process in which a uni-directional signal is obtained from a bi-directional signal. 'Power rectification' describes this process when it is used in obtaining a d.c. supply from an a.c. power source and is the principal topic of this chapter.

Figure 2.1(a) is a block schematic of a d.c. power supply unit (PSU) derived from the mains: Fig. 2.1(b) shows the overall function performed. At the input, v_I is a 240 V (RMS), 50 Hz sinusoid in Great Britain and, typically, a 110 V, 60 Hz sinusoid in the USA. The transformer provides d.c.-isolation for safety and, via the choice of primary/secondary turns ratio, a means of stepping-down or stepping-up the mains voltage without appreciable power loss. An additional feature of the transformer is that it provides a floating-output facility, which is considered further in Sections 2.3 and 2.4. At the output v_L is, typically, 5 V for logic and 15 V for analogue circuits.

The function of the boxes connected between the transformer and load is considered in the sections that follow. An evolutionary approach is adopted and, to aid understanding in the purely descriptive parts, the diodes used are assumed to have the ideal piecewise-linear characteristic of Fig. 1.7. This restriction is removed for calculations, in Section 2.2.

2.1 RECTIFICATION WITH A RESISTIVE LOAD

There are three cases to consider for a single-phase mains supply: half-wave; full-wave bi-phase; full-wave, bridge.

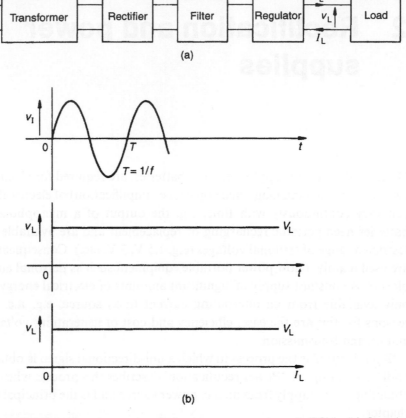

(a)

(b)

Fig. 2.1 (a) Block diagram of a mains-derived power supply unit (PSU). (b) Showing idealized characteristics

Half-wave (h.w.) rectification

The h.w. rectifier scheme is shown in Fig. 2.2. The load R is shown as a single resistor but it can represent the equivalent resistive loading of a piece of electronic equipment containing a number of components.

The mains input and secondary voltages v_I, v_S, respectively are given by,

$$v_I = V_{pm} \sin \omega t \tag{2.1a}$$

$$v_S = V_{sm} \sin \omega t \tag{2.1b}$$

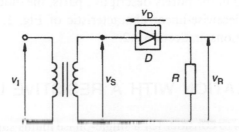

Fig. 2.2 Half-wave rectifier with resistive load

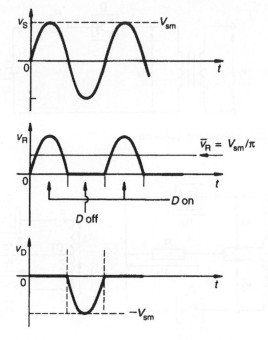

Fig. 2.3   Waveforms for Fig. 2.2

In the case considered, the peak primary voltage, V_{pm}, is greater than the peak secondary voltage, V_{sm}, by a factor n, the turns ratio of the transformer. Circuit operation is illustrated in the waveform sketches of Fig. 2.3. For $v_S > 0$, D (now called a 'rectifier diode') conducts and $v_R = v_S$. When $v_S < 0$, D is off and $v_R = 0$.

The mean value, $\bar{v}_R$, of the h.w.-rectified sinusoid is,

$$\bar{v}_R = V_{sm}/\pi \tag{2.2}$$

Rectifier diodes are usually listed with peak-inverse-voltage (PIV) ratings, rather than BV_R ratings, in component catalogues. In this case the PIV rating must exceed V_{sm}.

The h.w. rectifier gives a uni-directional, but not steady, output voltage. The associated uni-directional current pulses in the transformer give rise to d.c. magnetization of the core. This causes large magnetizing currents and waveform distortion, both of which are undesirable. Nevertheless, the circuit sometimes finds an application in battery charging. In Fig. 2.4 a resistor R_S is included to limit the peak charging current of the battery E.

Full-wave (f.w.) bi-phase rectification

Figure 2.5(a) shows the development of Fig. 2.2 into a f.w. rectifier. The transformer now has two secondary windings connected in series. Each winding is identical, in winding direction (as indicated in the dot notation) and the number of turns, to the secondary winding in Fig. 2.2. An equivalent of this f.w. scheme is

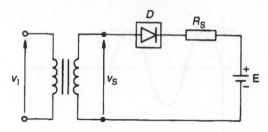

Fig. 2.4   Simple battery charging scheme

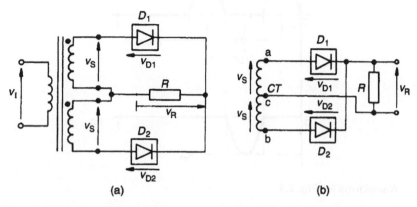

(a)   (b)

Fig. 2.5   Full-wave 'bi-phase' rectifier with resistive load. (a) Use of identical secondary windings. (b) Redrawn version of (a) but with centre-tapped transformer secondary

shown in Fig. 2.5(b). In this, the transformer has a single secondary winding, with twice the number of turns as that in Fig. 2.2, and a centre-tap (CT) connection.

For both configurations in Fig. 2.5, the anodes of D_1 and D_2 are driven by potentials that vary in opposite phase (hence 'bi-phase' circuit), as shown in Fig. 2.6. Hence,

$$\bar{v}_R = 2V_{sm}/\pi \tag{2.3}$$

The PIV rating of each diode must now exceed $2V_{sm}$.

The main advantages of f.w. compared with h.w. rectification are: no significant magnetization of the transformer core because secondary currents flow in opposite directions on alternate half cycles; and for a given $\bar{v}_R$, D_1 and D_2 each take half the mean current (alternatively, for a given diode current rating, the mean output voltage is doubled). These advantages are obtained at the cost of a more expensive transformer and diodes with a higher PIV rating.

The 'bridge' rectifier scheme

In this, f.w. rectification is achieved from a single-phase mains supply without the need for a centre tap on the transformer or for more than one secondary winding. The bridge comprises four diodes, a 'diode-quad', and it is important to

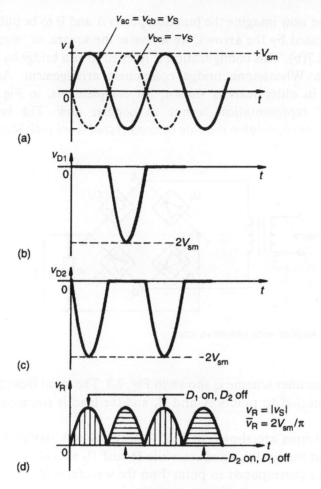

Fig. 2.6 Waveforms for Fig. 2.5. v_{ac}, v_{bc} constitute the two phases

remember how these are interconnected. As a memory-aid consider Fig. 2.7. Figure 2.7(a) shows two series-connected diodes, D_1, D_2 arranged in parallel with a similarly connected pair, D_3 and D_4. Note: at this stage all the diodes 'point the

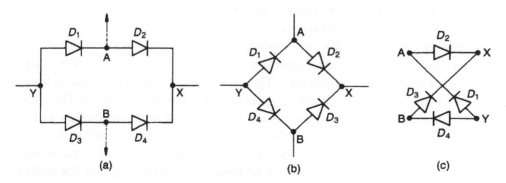

Fig. 2.7 (a) Diode bridge representation that emphasizes diode polarity. (b) Conventional pattern, derived from (a). (c) Older, 'figure-of-eight' pattern

same way'. If we now imagine the junction points A and B to be pulled out in the directions indicated by the arrows, we arrive at the square, or 'diamond', shape shown in Fig. 2.7(b). This configuration is referred to as a bridge by analogy with the well-known Wheatstone bridge component arrangement. An alternative drawing, used in older texts is shown, for completeness, in Fig. 2.7(c). This 'figure-of-eight' representation is not so popular now. The bridge can be purchased as an encapsulated module or constructed from individual diodes.

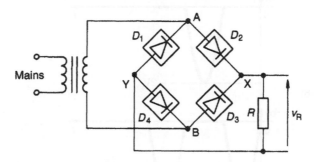

Fig. 2.8   Bridge rectifier with resistive load

A basic f.w. rectifier scheme is shown in Fig. 2.8. The input from the secondary winding is connected to points A and B, and the load is connected between X and Y.

Circuit waveforms are shown in Fig. 2.9. For $v_S > 0$, diodes D_2 and D_4, in opposite arms of the bridge, conduct while D_1 and D_3 are cut off. Point J' on the waveform for v_R corresponds to point J on the waveform for v_S. An equivalent circuit for instantaneous conditions at $t = t_1$, is shown boxed. The arrows show the direction of current flow around the circuit.

Operation for $v_S < 0$ closely mirrors that for $v_S > 0$. In this case D_1 and D_3 conduct, while D_2 and D_4 are cut off, and point K' on the waveform for v_R corresponds to point K on the waveform for v_S. This follows from the boxed instantaneous equivalent circuit, in which the direction of current flow is shown by the broken-line arrows. As in the case of the bi-phase circuit, Equation (2.3) characterizes the mean output voltage.

Comparing the bridge with the bi-phase rectifier schemes, for a given v_R, we reach the following conclusions regarding the transformer and diodes. The bridge circuit uses a lower cost transformer. This comes about because it requires fewer secondary turns, no centre-tap arrangement and a lower current rating. The latter property arises because the current flows continuously in the secondary rather than half the time in each winding. The PIV rating of the diodes used is only one half that for the bi-phase circuit. The use of four diodes instead of two does not increase the cost significantly. Because of these major advantages of the bridge scheme it is normally used in preference to the bi-phase scheme, as in the circuits considered in the remainder of this chapter.

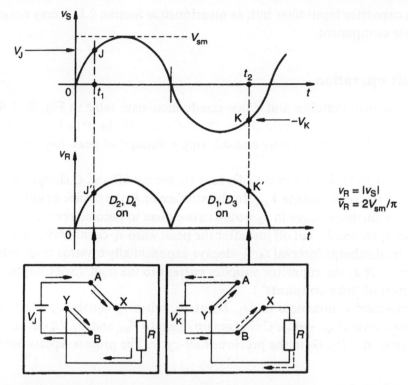

$v_R = |v_S|$

$\bar{v}_R = 2V_{sm}/\pi$

Fig. 2.9 Waveforms for Fig. 2.8. Boxed sections show instantaneous equivalent circuits at $t = t_1$, $t = t_2$

2.2 USE OF A RESERVOIR CAPACITOR

At its simplest, the filter box in Fig. 2.1 comprises a single component, a 'reservoir' capacitor C (usually electrolytic), connected in parallel with the load to give a smoothed output. The description 'capacitor input filter' is used for this because the diode rectifier output section 'looks into' a capacitor, as shown in Fig. 2.10. The load voltage is labelled v_C rather than v_R because C is always present

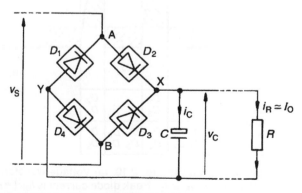

Fig. 2.10 Full-wave rectifier with capacitor-input filter

with a capacitor input filter but, as mentioned in Section 2.1, R may not exist as a separate component.

Circuit operation

The waveform sketches and diode-conduction-state table in Fig. 2.11 illustrate circuit operation. Waveform $|v_C|$ is that of $|v_S|$ with the gaps between the peaks almost filled in. There is only a small 'ripple voltage' of peak-to-peak magnitude V_r.

At point P_1 in the interval t_c, D_2 and D_4 are cut off and C charges up through D_1, D_3 to the peak voltage V_{sm}. In the absence of R, the diodes would then cut off because a further change in v_S would cause them to become reverse-biased. With R present, D_1 and D_3 cut off just after the peak when v_C cannot fall as fast as v_S. At P_2, in the discharge interval t_d, v_C decays exponentially towards zero with a time-constant CR as the capacitor supplies current to the load. (*See* Chapter 10 for a discussion of 'time-constants'.)

Conduction commences in D_2, D_4 at P_3, when v_S equals v_C, which has now fallen to a level $(V_{sm} - V_r)$. C charges up again to V_{sm} and D_2, D_4 cut off just after the peak as did D_1, D_3 in the previous half-cycle. The process repeats continually.

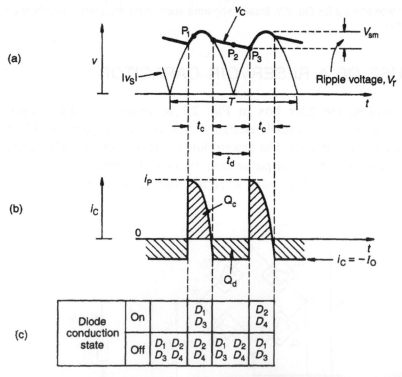

(a)

(b)

(c)

Fig. 2.11   Showing operation of circuit in Fig. 2.10. (a) Voltage waveforms. (b) Capacitor current waveform, for practical case, $i_R \equiv I_0$. Peak diode current is I_{DP} ($= I_P + I_0$). (c) Diode conduction table

Current flow in opposite pairs of the bridge diodes is thus limited to short intervals, t_c, in the vicinity of the peaks of $|v_S|$ and during these intervals a charge $Q_c = CV_r$ is supplied to C. The discharge current of C is approximately constant at $I_O = V_{sm}/R$ for the practical case $CR \gg (T/2)$, depicted in Fig. 2.11. Then the charge lost by C in the interval t_d is $Q_d = I_O t_d$. After initial switch-on, a state of equilibrium exists when $Q_c = Q_d$. The peak capacitor current, i_P, (and diode peak current, $I_{DP} = i_P + I_O$) is large compared with I_O because $t_c \ll t_d$.

Bridge circuit calculations

In a more detailed analysis of rectifier circuit operation, Fig. 2.10 requires modification in two respects. First, allowance must be made for the finite voltage drop across a conducting diode. A useful approximation to practical performance is achieved by assuming a constant forward voltage drop, δ, across a bridge diode when it conducts (*see* Chapter 1).

Second, it is traditional and convenient to plot circuit voltages as a function of phase angle θ ($= \omega t$) in radians, rather than time, t, because the transformer secondary voltage is expressed as $v_S = V_{sm} \sin \omega t$. Since $\omega = 2\pi f = 2\pi/T$, the relationship for converting from θ to t is $t = T(\theta/2\pi)$, so plots can be re-scaled, if desired, or drawn with dual horizontal scales.

These modifications are incorporated in Fig. 2.12, in which $|v'_S|$ represents the effective magnitude of the transformer secondary voltage. Thus,

$|v'_S| = 0$ for $v_S < 2\delta$ and $|v'_S| = v_S - 2\delta$ for $v_S > 2\delta$

Parameters of interest in calculations are: θ_1 = 'cut-in' angle, and associated peak diode current I_{DP}; θ_2 = 'cut-out' angle; θ_c = 'conduction' angle = $(\theta_2 - \theta_1)$; V_r = peak-to-peak ripple voltage; V_{rm} = theoretical maximum value of V_r; V_O = mean output voltage. θ_1, θ_2, V_r are all inter-related, a determination of one of them requiring a knowledge of the others. They can be found from graphical plots or, as below, by approximate methods. More accurate results can be obtained for V_r by the iterative solution of two simultaneous equations (*see*

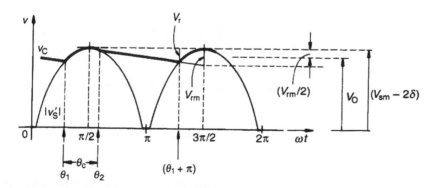

Fig. 2.12   A version of Fig. 2.11(a) used in analysis

Substituting V_{rm} for V_r and simplifying,

$$\sin \theta_1 = 1 - (V_{rm}/V_{sm}) \tag{2.16}$$

Referring to Equation (2.13), i has its maximum value I_{DP} at $\omega t = \theta_1$, when $\cos \omega t$ has its largest value. Thus,

$$I_{DP} = (\omega C V_{sm} \cos \theta_1) + I_O \tag{2.17}$$

The mean current in each diode is $(I_O/2)$. Finally, since $\theta_2 \simeq (\pi/2)$,

$$\theta_c \simeq (\pi/2) - \theta_1 \tag{2.18}$$

Example 2.1

A transformer fed from the 240 V, 50 Hz mains has a primary/secondary turns ratio of 10:1. The secondary winding feeds a bridge rectifier system, with a capacitor input filter, that supplies a load current of 100 mA. Determine a value for the reservoir capacitor if the peak-to-peak ripple voltage is not to exceed 0.5 V and calculate the required peak current and PIV ratings of the bridge diodes. (Neglect transformer winding resistance and assume the drop across a conducting diode is 1 V.)

Solution

240 V is an RMS value: $V_{pm} = 240\sqrt{2}$
Hence, $V_{sm} = 240\sqrt{2}/10 = 34$ V, $(V_{sm} - 2\delta) = 32$ V
Since $V_r \, (= 0.5$ V$) \ll 32$ V we can assume $V_r = V_{rm} = I_O T/2C$, or $C = I_O T/2V_r$
Substituting $V_r = 0.5$ V, $I_O = 100$ mA, $T = 1/f = 20$ ms, gives,

$$C = 2 \text{ mF } (= 2000 \ \mu\text{F})$$

Using Equation (2.16),

$$\sin \theta_1 = 1 - (0.5/32) = 0.985$$

$$\therefore \theta_1 = 80.16° \text{ and } \cos \theta_1 = 0.171$$

Substituting in Equation (2.17),

$$I_{DP} = [(2\pi \times 50 \times 2 \times 10^{-3} \times 34 \times 0.171) + 0.1] \text{ A}$$

$$\therefore I_{DP} = 3.75 \text{ A}$$

Thus $I_{DP} \gg I_O$, as mentioned in Section 2.2. The mean current in each diode is 50 mA. Conservative ratings for I_{DP} and PIV are, respectively, 5 A and 50 V.

The capacitor input filter, alone, does not provide the degree of smoothing normally required in equipment design. The effect of ripple can be reduced by the addition of an LC filter as shown in Fig. 2.14 (*see* Problem 8), but the disadvantages of using an inductor (or 'choke') in this context include: d.c.

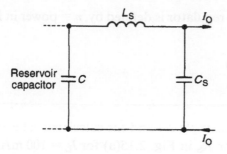

Fig. 2.14 L_S, C_S comprise a simple smoothing filter

resistance; requirement to have a high inductance for load currents up to 1 A, or more; weight; and cost. The use of an inductor can often be avoided by employing a regulator.

2.3 USE OF A REGULATOR

Two basic types of regulator are the zener diode shunt regulator and the IC regulator.

Zener shunt regulator

This is shown in Fig. 2.15(a). A d.c. equivalent circuit, that incorporates Fig. 2.13(b) for the rectifier and Fig. 1.23(b) for the zener diode, is shown in Fig. 2.15(b). Following the analysis given in Section 1.2, V_Z is obtained from,

$$(V_{sm} - 2\delta - V_Z)/(R_S + r_O) = [(V_Z - V_{ZS})/r_z] + I_{ZS} + I_L \qquad (2.19)$$

The peak-to-peak ripple voltage, V_{rL}, in the load follows from the use of the circuit in Fig. 2.15(c).

$$V_{rL} = V_r r_z/(r_z + R_S) \qquad (2.20)$$

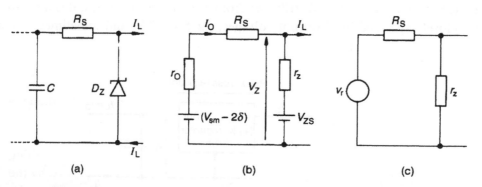

Fig. 2.15 (a) Use of zener regulator with capacitor–input filter. (b) D.C. equivalent circuit of (a). (c) Circuit for calculating ripple voltage in load

The efficiency, η, of the regulator is defined by, η = power in load/power supplied. Hence,

$$\eta = I_L V_Z / I_O (V_{sm} - 2\delta) \qquad (2.21)$$

Example 2.2

Use the data given in Example 2.1 and the calculated value of reservoir capacitor to determine a value for R_S in Fig. 2.15(a) for $I_L = 100$ mA. Calculate, also, the peak-to-peak ripple voltage across the load. Data for D_Z is $V_{ZS} = 15$ V, $r_z = 20\,\Omega$ at $I_{ZS} = 15$ mA.

Solution

$$r_O = 1/4fC = 1/(4 \times 50 \times 2 \times 10^{-3}) = 2.5\,\Omega$$

Assume $I_Z = I_{ZS} = 15$ mA. Then, $V_Z = V_{ZS}$, $r_z = 20\,\Omega$ and $I_O = 115$ mA = 0.115 A. Rearranging Equation (2.19) and substituting the given data yields,

$$R_S(\Omega) = [(34 - 2 - 15)/0.115] - 2.5 = 145$$

A series combination of a 43 Ω, 0.5 W resistor and a 100 Ω, 2 W resistor meets the requirement.

$$V_r = 2r_O I_O = 0.5\,\Omega \times 115\text{ mA} = 575\text{ mV}$$

From Equation (2.20), V_{rL} (mV) = 575 $\times$ 20/163, i.e, $V_{rL} \simeq 71$ mV.

IC regulators

A problem with the zener regulator is the poor efficiency ($\simeq$40% in Example 2.2) that accompanies ripple reduction. This is not necessarily the case with an IC regulator.

The IC regulator shown in Fig. 2.16 is a basic block in PSU design. It has the following properties: low cost, thus facilitating its use on an individual printed circuit board if 'raw' d.c. is distributed from a central rectifier unit; it is available with closely specified fixed output voltages, e.g. 5 V, 12 V, and sometimes with a

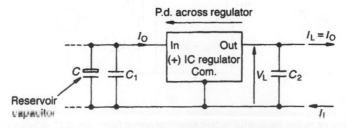

Fig. 2.16   A positive voltage regulator, showing directions of current flow and voltage drop

programmable output arrangement; $TCV_L \sim 1$ mV/°C; wide input voltage range; low output resistance (typically, < 50 mΩ); high ripple-reduction factor; and output short-circuit protection.

Capacitors such as C_1, C_2 are recommended by the manufacturer, C_1 to minimize the effect of any unwanted signals 'picked-up' on possibly-long input leads (i.e. C_1 is a 'decoupling' capacitor), and C_2 to improve transient response.

Figure 2.16 shows a positive regulator. This is one designed to operate with currents flowing in the directions shown. These directions are reversed for a negative regulator. Suppose the load connected across C_2 in Fig. 2.16 is a resistor, R. The resulting load current is $I_L = (V_L/R)$. Now, $I_O = (I_L + I_X)$ where $I_X (\ll I_L)$ accounts for small currents flowing into the regulator itself. I_X is sensibly constant as V_O varies. Hence, I_O is constant as the potential difference across the reservoir capacitor changes. This validates the assumption of constant I_O made in Section 2.2.

In Fig. 2.16, V_L is 'floating' in the same sense that the battery supply for powering a pocket torch is floating. Thus the potential difference between the terminals is known, but the potential difference between either of these and a fixed potential such as 'earth' (*see* next section) is not. Either terminal of C_2 can be connected to a reference voltage V_{ref}.

Figure 2.17 shows a dual-output PSU employing a bridge rectifier assembly and two regulators, one positive and the other negative. In this case a centre-tapped secondary winding is used and CT is 'earthed'.

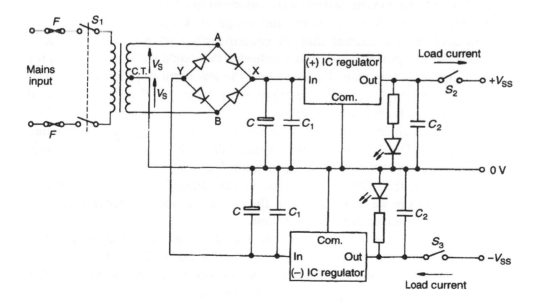

Fig. 2.17   A dual-polarity PSU using positive and negative regulators

2.4   EARTH AND 'EARTHING' ('GROUNDING')

In dealing with electronic circuits, in general, and regulators and PSUs, in particular, it is important to distinguish between 'true-earth' and 'chassis-earth'.

A 'true-earth', symbolized in Fig. 2.18(a), is a node that is ultimately connected, perhaps via a lead in the mains plug and socket, to a metal plate (or pipe) buried in the subsoil of planet Earth. Because of the low resistance of the soil and large capacitance of the Earth, this provides a convenient reference for 'zero volts'. An analogy is the use of mean sea level to specify the height of mountain peaks and ocean beds. Even if an amount of water that would fill the United Nations building in New York were to be dumped in, or removed from, the sea there would be no perceptible change in sea level.

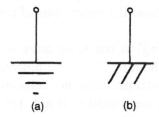

(a)   (b)

Fig. 2.18   'Earth' symbols: (a) 'true-earth'; (b) 'chassis-earth'

A 'chassis-earth', or 'chassis-ground' (symbolized in Fig. 2.18(b)), is a 'common' or reference node against which all others in a circuit, or system, are reckoned. It may be regarded as a 'local' earth.

Whether or not a chassis-earth is also connected to a true-earth depends on the circumstances of equipment design and usage. In stationary equipment it is desirable and usual to ensure that the chassis-earth is connected to true-earth. This stops the equipment floating to a potential that could make it hazardous for a human operator to touch. Furthermore, the low earth impedance minimizes electrostatic pick-up from other equipment, particularly if an earthed metal enclosure surrounds sensitive circuitry.

In portable equipment the chassis-earth is normally, unavoidably, isolated from true-earth. Thus, with a car the metal of the bodywork provides a chassis-earth, for electrical and electronic equipment, but it is insulated from the true-earth of a road by the rubber of the tyres. The disconnection between chassis and true-earth is even more pronounced in the case of circuits in a space rocket, in transit to another planet in the solar system.

In the literature the two symbols in Fig. 2.18 are sometimes used interchangeably. For standardization purposes the symbol in Fig. 2.18(b) is used, from now on, in place of a line marked '0 V'. However, the remarks made in this section should be borne in mind in a specific application.

2.5 RELATED CIRCUITS

Some circuit schemes that exploit the charge storage feature of the capacitor input filter are considered briefly here. Figure 2.19 shows a basic pulse-stretcher circuit, a modified form of which is used in nuclear pulse instrumentation.

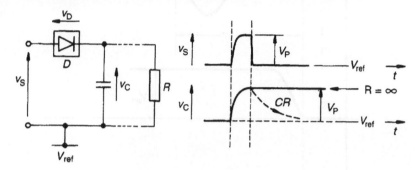

Fig. 2.19   A simple pulse-stretcher and its waveforms

During the pulse, C charges up through D to the peak voltage V_P and retains this, thereafter, in the absence of R. With R present (the practical case) v_C decays exponentially to zero with time-constant CR. The circuit is also the basis of a peak voltage detector.

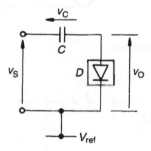

Fig. 2.20   A basic clamping circuit (a re-arrangement of the pulse stretcher)

A 'clamping' circuit is shown in Fig. 2.20. Versions of this are used in instrument probes and in the amplitude-control of sinusoidal oscillators. Circuit operation for a sinusoidal input waveform is shown in Fig. 2.21. In the first positive half-cycle of input, C charges up through D to a peak value V_{sm} and stores this voltage. Then,

$$v_O = v_D = v_S - V_{sm} = V_{sm} \sin \omega t - V_{sm}$$

The peak level of the output is, thus, 'clamped' to a reference level V_{ref}, often zero.

A final example is the voltage-doubler circuit of Fig. 2.22. The upper reservoir capacitor charges up to V_{sm} when $v_S > 0$ and the lower capacitor to the same value when $v_S < 0$.

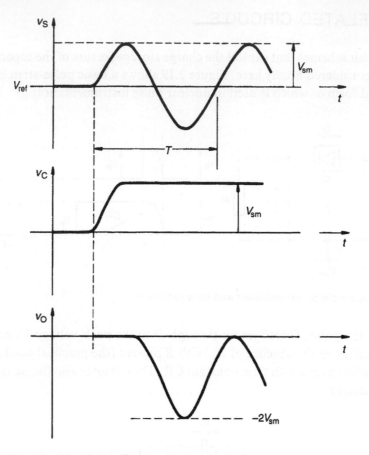

Fig. 2.21   Clamping circuit waveforms

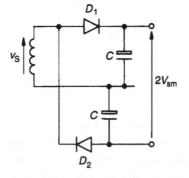

Fig. 2.22   A voltage-doubling circuit

2.6   SELF ASSESSMENT TEST

1 What are the advantages of a f.w. bi-phase rectifier compared with a h.w. rectifier?

2 What happens if one of the diodes in a bi-phase rectifier fails and becomes an open circuit?

3 In what respects is a bridge rectifier scheme superior to one using f.w. bi-phase rectification?

4 Why does the diode current flow in short pulses when a capacitor input filter is used?

5 What happens to the diode peak current if the load current on a capacitor input filter increases but the ripple voltage is required to remain unchanged?

6 What effect does increased supply frequency have on rectifier circuit design?

7 What are the characteristics of an IC voltage regulator?

2.7 PROBLEMS

1 Show that the cut-in angle, α, cut-out angle $(\pi - \alpha)$ and mean charging current $\bar{I}$, for the circuit of Fig. 2.4 are given by,

$$\alpha = \sin^{-1}(E/V_{sm})$$

$$\bar{I} = (1/2\pi R_S) \int_{\alpha}^{(\pi-\alpha)} (V_{sm} \sin \omega t - E)d(\omega t)$$

2 Perform the integration in Problem 1 and show that,

$$\bar{I} = [(V_{sm} \cos \alpha)/\pi R_S] - [(\pi - 2\alpha)E/2\pi R_S]$$

3 A moving coil milliammeter is connected in series with the resistor R in Fig. 2.8. Calculate R for a meter indication of 1 mA, mean, if the RMS value of v_S is 240 V (ignore meter resistance).

4 Calculate an upper limit for percentage regulation for the assumptions made in Section 2.2.

5 Derive Equation (2.5) by equating the energy lost by the capacitor in the discharge period to the energy supplied to the load.

In fact, the name transistor (*transfer resistor*) arose because it was realized, at the outset, that power gain would be achieved by transferring a current from a low resistance control circuit to a high resistance load circuit.

In circuit theory, devices which exhibit power gain are called 'active' devices. 'Passive' devices do not give power gain, they only dissipate energy (e.g. resistor, diode) or store it (e.g. capacitor, inductor).

The two polarities, P and N, of semiconductor material give rise to two types of BJT structure, the NPN device shown schematically in Fig. 3.1(a) and the PNP device, shown in Fig. 3.2(a). (A more realistic structure for an NPN device is mentioned in Section 3.5). The three layers are sometimes referred to as a 'sandwich' but, of course, they are formed within a single crystal of material, i.e., they are not three separate layers joined together.

In Figs 3.1(b), 3.2(b) the arrow inside the circuit symbol is associated with the emitter terminal. The direction of the arrow, from P material to N material, shows whether the device is NPN or PNP. The reference directions for conventional current flow of the terminal currents, I_C, I_B, I_E, are shown by their associated arrows. These currents are positive when BJTs are used for amplification.

Each BJT type may be Si or Ge, but Si is nearly always used because of operational and manufacturing reasons: Si devices can operate at higher voltages

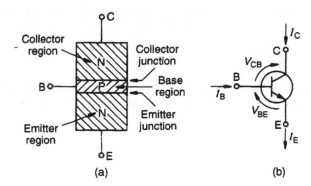

(a)   (b)

Fig. 3.1   (a) Schematic cross-section of an NPN transistor. (b) Circuit symbol with current and voltage lettering

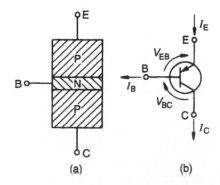

(a)   (b)

Fig. 3.2   (a) Cross-section of a PNP transistor. (b) Annotated circuit symbol and higher temperatures than their Ge counterparts; stable oxide layers can be grown on Si surfaces during the manufacture of both discrete devices and monolithic ICs. These layers give 'clean' junctions in the case of discrete devices and provide other useful functions, e.g. insulation, in the case of IC structures. Note: Si NPN devices are normally used in preference to PNP types because they are easier to make and have superior terminal characteristics, so the emphasis henceforth is on Si NPN transistors.

The letter often used for a BJT is 'Q', with appropriate subscript(s) if required: the designations 'T', 'Tr' can be confused with temperature and transformer respectively, if circuit action involves these. (As with the diode, Q also refers to a quiescent point but the context makes clear what the letter represents.)

3.2 REGIONS OF OPERATION

Figure 3.3 indicates possible operating 'regions' or 'modes' of an NPN device: the diagram also applies to PNP devices if V_{CB} is replaced by V_{BC} and V_{BE} by V_{EB}. As there are two PN junctions each of which may be either forward-biased or reverse-biased, there are four operating regions.

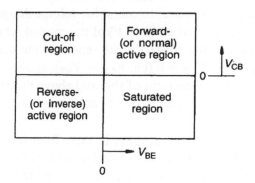

Fig. 3.3   Operating modes of an NPN transistor. For a PNP device, replace V_{CB} by V_{BC} and V_{BE} by V_{EB}

The forward-active, or normal-active, region ($V_{BE} > 0$, $V_{CB} > 0$) is used for analogue applications. The cut-off region ($V_{BE} < 0$, $V_{CB} > 0$) and the saturated region ($V_{BE} > 0$, $V_{CB} < 0$) correspond, respectively, to the open and closed states of a BJT switch, used in digital applications. The reverse-active, or inverse-active, region ($V_{BE} < 0$, $V_{CB} < 0$) has no known useful applications.

A crude analogy for these regions is provided by the states of motion of a motor car. In the forward-active region the car is in gear and moving forwards: in the reverse-active region it is moving backwards. A parked car corresponds to cut-off: a car in gear but restrained from moving by a massive spring connected to a fender corresponds to saturation. The BJT used as a switch is considered in Chapter 10. Up to that point, operation in the forward-active mode, only, is discussed. For an explanation of the physical electronics of device operation in this region the reader is referred to the literature (*see*, e.g. Muller and Kamins, 1977).

3.3 THE 'IDEAL LOGARITHMIC' (OR EM) TRANSISTOR MODEL

Figure 3.4 serves to define the d.c. properties of an 'ideal logarithmic' NPN transistor. This is the simplest description that approximates the behaviour of a practical device.

The circuit representation in column 1 is a version of the model due to Ebers and Moll (EM), the first workers to give a systematic description of d.c. performance. The base-emitter circuit of the device is modelled by an ideal logarithmic diode, the collector-emitter circuit by an ideal current source. The current source may be regarded as controlled by either the voltage drop across the diode or to the current flowing in it. This gives rise to the descriptions voltage-controlled current source (VCCS) and current-controlled current source (CCCS) for the BJT.

The sets of characteristics in Fig. 3.4 are not exhaustive, but most relevant to practical conditions. The characteristics are classified according to the following criteria: the common, or reference, terminal for voltage designation; the terminal-pair to which they refer; the nature (I or V) of the control, or drive, signal. In each of the plots in column 3, the shaded areas represent the saturation region. The boundary curve for this is $V_{CB} = 0$ (i.e. $V_{CE} = V_{BE}$), shown dotted. The forward-active mode is in the area to the right of this and above the axis line $I_C = 0$, which is in the cut-off region. The characteristics in columns 4 and 5 refer to the forward-active region.

Rows (a), (b), (c) are considered, now, in turn. Row (a) depicts the common-emitter (CE) characteristics. These are so-called because the emitter is regarded as the reference terminal: V_{BE} and V_{CE} are the relevant terminal voltages. In this case, the BJT is regarded as a VCCS so the CE output (or 'collector') characteristics in column 3 display the family of curves $I_C = f(V_{CE})$ for set values of V_{BE}. The CE input characteristic (column 4) is the plot of $I_B = f(V_{BE})$, which is a single curve for $V_{CB} \geqslant 0$.

The CE transfer, or mutual, characteristic (column 5) is a plot of I_C vs V_{BE}. In this case log-linear scales are used. The functional relationships for $V_{BE} > 100$ mV and $V_{CB} = 0$ are,

$$I_C = I_S e^{V_{BE}/V_T} = A_J J_S e^{V_{BE}/V_T} \tag{3.1}$$

$$I_B = (I_S/\beta_0) e^{V_{BE}/V_T} = (I_C/\beta_0) \tag{3.2}$$

$$I_E = I_C + I_B \tag{3.3}$$

Equation (3.3) is a consequence of Kirchhoff's Current Law (KCL) and is not a consequence of BJT action.

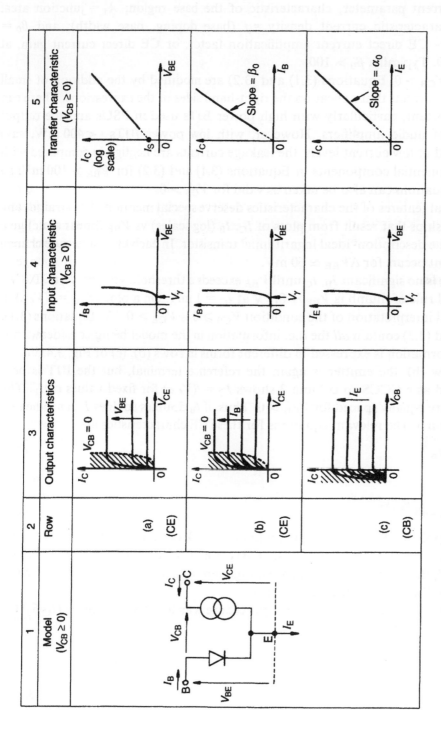

Fig. 3.4   D.C. terminal characteristics of an idealized NPN transistor. (Small leakage currents neglected.) Curves to the right of saturation boundary line $V_{CB} = 0$ correspond to the Ebers–Moll (EM) 'ideal logarithmic' model

In Equations (3.1) and (3.2): V_T = thermal voltage, introduced in Chapter 1; I_S = current parameter, characteristic of the base region; A_J = junction area; J_S = characteristic current density = f (base doping, base width); and β_0 = (I_C/I_B) = CE direct current amplification factor, or CE direct current gain, at $V_{CB} = 0$. Typically, $\beta_0 > 100$.

For $V_{CB} > 0$, Equations (3.1) and (3.2) are modified by the addition of small leakage currents that appear on the right-hand sides of the expressions. These can be important, particularly with high power BJTs used in PSUs and the output stages of audio amplifiers. However, with low power BJTs (<400 mW, say) operated at low current levels, the leakage currents are negligible compared with the exponential components in Equations (3.1) and (3.2) for $V_{BE} > 100$ mV, so these equations can also be taken as valid for $V_{CB} > 0$.

Several features of the characteristics deserve special mention. The straight line relationships that result from plots of I_C, I_B (log scales) vs V_{BE} (linear scale) give rise to the description 'ideal logarithmic' transistor. In each case, a decade change in current occurs for $\Delta V_{BE} \simeq 60$ mV.

There is no significant I_B, I_C until V_{BE} exceeds a threshold voltage, V_γ ($\simeq 0.6$ V). A useful rule of thumb is $V_{BE} \sim 0.7$ V at $I_C = 1$ mA. On a plot of $I_C = f(V_{CE})$, a practical interpretation of the condition $V_{CB} \geqslant 0$ is $V_{CE} \geqslant 0.7$ V. Equations (3.1), (3.2) and (3.3) contain *all* the d.c. information in the model being considered, but that information is expressed in different forms in rows (b), (c) of Fig. 3.4.

In row (b), the emitter is again the reference terminal, but the BJT is now regarded as a CCCS so column 3 shows $I_C = f(V_{CE})$ for fixed values of I_B. The curves are equally spaced for $V_{CB} > 0$. Thus, if I_B doubles so does I_C, as indicated in column 5. The relevant equations for these CE characteristics are,

$$I_C = \beta_0 I_B \tag{3.4}$$

and,

$$I_B = (I_S/\beta_0)e^{V_{BE}/V_T} \tag{3.5}$$

These apply for $V_{CB} = 0$ and, again ignoring leakage currents, for $V_{CB} > 0$ also. Row (c) refers to the common-base (CB) connection. As the base is the reference terminal, the appropriate terminal voltages are V_{CB} and V_{BE}. The BJT is viewed as a CCCS with input I_E and output I_C.

To find $I_C = f(I_E)$ at $V_{CB} = 0$ the procedure is as follows. Substitute (I_C/β_0) for I_B in Equation (3.3),

$$I_E = I_C + (I_C/\beta_0)$$

or,

$$I_C = \beta_0 I_E/(1 + \beta_0)$$

Let,

$$\alpha_0 = \beta_0/(1 + \beta_0) \tag{3.6}$$

Then,

$$I_C = \alpha_0 I_E \tag{3.7}$$

α_0 = CB direct current amplification factor, or CB direct current gain, at $V_{CB} = 0$. Like β_0, α_0 is independent of I_C. Furthermore, $\alpha_0 > 0.99$ for $\beta_0 > 100$. The curve $I_E = f(V_{BE})$ is found by substituting $I_C = \alpha_0 I_E$ in Equation (3.1). The result, after transposition is,

$$I_E = (I_C/\alpha_0)e^{V_{BE}/V_T} \tag{3.8}$$

Equations (3.6), (3.7) and (3.8) also apply for $V_{CB} > 0$ if, as assumed above, leakage currents are ignored.

Table 3.1 summarizes the relationships between I_C, I_B, I_E and facilitates a conversion between them. The parameter β_0 is used in the relationships but the Table entries can also be expressed in terms of α_0 (*see* Problem 1).

Table 3.1   Terminal current conversion chart

To get this → from this ↓	multiply by the factor in the square	I_B	I_C	I_E
I_B		1	β_0	$(1 + \beta_0)$
I_C		$1/\beta_0$	1	$1 + (1/\beta_0)$
I_E		$1/(\beta_0 + 1)$	$\beta/(1 + \beta_0)$	1

In the forward-active mode the EM model of the BJT, or a simplified version of it, is suitable for pencil-and-paper calculations of d.c. conditions in BJT circuits, as in Section 3.6. However, the model requires modification for computer calculations and for deriving an incremental model, used in Chapter 4 for amplifier applications. An improved d.c. model is considered next.

3.4   THE EME (EBERS–MOLL–EARLY) D.C. MODEL

The main limitation of the model of the previous section is that it does not take into account the finite slope that occurs on the output characteristics of real BJTs, in the forward-active region. That slope is allowed for in, what is called here, the EME model. The third letter in the description refers to Early, the engineer who first investigated the effect quantitatively.

In the EME model, the CE output characteristics, for $V_{CB} > 0$, are approximated by their tangents at $V_{CB} = 0$. When extrapolated back, they appear to radiate from a common axis point $V_{CE} = -V_A$. V_A is known as the 'Early voltage'. In the model, an increase in V_{CB} causes an increase in I_C but, surprisingly perhaps, *not* I_B. The relevant equations are,

where,

$$\beta = (I_C/I_B) = \beta_0[1 + (V_{CE}/V_A)] \tag{3.15}$$

β is the CE d.c. current gain for $V_{CB} > 0$ (i.e. $V_{CE} > 0.7$ V).

For row (c), the common-base characteristics in column 3 appear to originate from a common intersection point $V_{CB} = -(\beta_0 + 1)V_A \simeq -\beta_0 V_A$

From Equations (3.9), (3.10) and (3.11),

$$I_E = I_S[1 + (1/\beta_0) + (V_{CB}/V_A)]e^{V_{BE}/V_T} \tag{3.16}$$

This describes the family $I_E = f(V_{BE})$ in column 4.

The slopes of the transfer characteristics in row (c), column 5, are given (*see* Problem 2) by,

$$\alpha = (I_C/I_E) \simeq \alpha_0[1 + (V_{CB}/\beta_0 V_A)] \tag{3.17}$$

α is the CB direct current gain for $V_{CB} > 0$ and is independent of I_E.

Figure 3.3 and all the equations describing NPN performance apply equally well to PNP transistors, for the current directions shown in Fig. 3.2, if V_{EB} is substituted for V_{BE} and V_{BC} ($\geqslant 0$) for V_{CB}. In concluding this section, note that the load line technique of Chapter 1 is applicable to both the base and collector circuits of the BJT.

3.5 PRACTICAL POINTS

Structure

Figure 3.7 is a more realistic representation of the cross-section of a discrete NPN transistor than Fig. 3.1. N^+ refers to a richly doped N region. A vertical slice between the dotted lines ff', gg' defines the active area of the base region. The

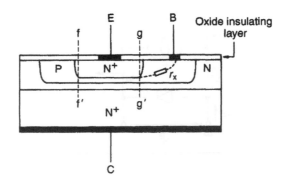

Fig. 3.7 More realistic cross-section of a discrete NPN transistor showing the existence of 'extrinsic base resistance', r_x. N^+ denotes richly doped N region

resistance between the base contact, B, and this active region is modelled by a resistor r_x (shown dotted), variously known as the 'extrinsic base resistance', 'base bulk resistance' or 'base spreading resistance'.

It is not necessary to include r_x in the d.c. models of low power BJTs because the potential difference across the base resistance is negligible compared with observed values of base–emitter voltage. However, it is necessary to take r_x into account in BJT incremental models (Section 4.3).

Characteristics

Equation (3.1) is obeyed precisely over a collector current range of some eight decades below 5 mA, but Equation (3.2) is valid over a much narrower range, i.e. β_0 is current dependent. Some authors prefer the symbols h_{FB}, h_{FE} for α, β, respectively, but this entails the use of two subscripts which might be confused.

Typical manufacturing tolerances, for an NPN transistor, yield a 3:1 spread on β and a ± 30 mV spread on V_{BE}, at a given I_C ($A \pm 50$ mV spread in V_{BE} is a safe allowance in circuit design). However, similar devices made on the same IC chip normally show a β-match of $\pm 5\%$ and a V_{BE}-match of ± 1 mV. V_A is not usually specified: a default value is 100 V. In general: $TCV_{BE} \sim -2$ mV/°C; $TCI_S \sim 16\%$ per °C; $TC\beta \sim 1\%$ per °C.

Ratings

A number of breakdown voltages exist. At an introductory level the most important are BV_{CEO}, BV_{CBO} ($> BV_{CEO}$) and BV_{BR}, none of which must be exceeded. BV_{CEO} (*see* Fig. 3.8) is the collector–emitter breakdown voltage measured when the base is open circuit. Similarly, BV_{CBO} is the breakdown

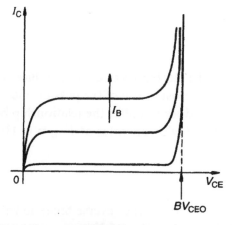

Fig. 3.8 Breakdown voltage BV_{CEO} is applicable to common–emitter operation

voltage between collector and base with the emitter open circuit. BV_{BR} refers to the base–emitter breakdown voltage.

The collector dissipation, P_C, of a BJT under d.c. conditions is given by $P_C = I_C V_{CE}$, the base dissipation by $P_B = I_B V_{BE}$. Since $V_{BE} \leqslant V_{CE}$, in the forward-active region, and $I_B \leqslant (I_C/100)$, it follows that $P_B \leqslant (P_C/100)$. Data sheet information on total device dissipation can thus be taken as applying solely to P_C. The derating graph, for a BC107, shown in Fig. 3.9 is based on data sheet information and is constructed in the same way as that for a diode.

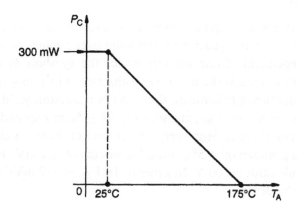

Fig. 3.9   Collector dissipation derating graph for the BC107

BJT classification

Transistor outline (TO) packages range from plastic and metal cans for low power BJTs to stud-mounted packages for high power devices such as the 2N3055, which is capable of dissipating some tens of watts at $T_A = 25°C$. A coding chart is shown in Table 3.2.

Related Devices

Strapping-together two of the electrodes of a BJT produces two related devices, a diode and a zener diode. The most useful of the possible diodes is formed by using a C–B strap (Fig. 3.10). This does not alter the relationship between I_B, I_C, I_E, that exists for the BJT since the device operates at $V_{CB} = 0$. However, the terminal current relationship becomes,

$$I = (I_S/\alpha_0)e^{V_D/V_T} \tag{3.18}$$

Operating the diode of Fig. 3.10 in the reverse breakdown region produces the zener diode in Fig. 3.11. Typically, $V_Z \sim 6$ V.

Table 3.2 BJT classification chart

Name	Details
American	JEDEC

2N × × × ×

Two junctions   Registration number

European	(Proelectron) 5-symbol α-numeric code.
	Two numbers = Industrial & professional use.
	Three numbers = Consumer use.

────────────────────────── Registration number

C = A.F. low power: D = A.F. power
A = Ge   F = H.F. low power: L = H.F. power
B = Si   S = Low power switch; U = power switch
C = GaAs

CV (common valve)	Military Code in U.K.
'In-house' code	Manufacturers own code

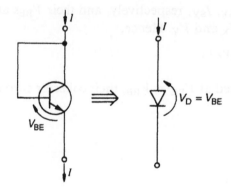

Fig. 3.10   Diode formed by a collector–base strap

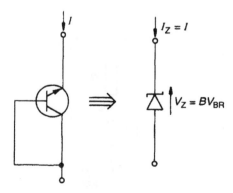

Fig. 3.11 Use of strapped device as a zener diode

3.6   BIASING THE BJT

This means choosing a network in which to embed the BJT to obtain predictable quiescent values of I_C and V_{CE}. (Note: An extra subscript Q can be used to distinguish between the quiescent and general values of d.c. variables but is not necessary here since quiescent values, only, are being considered.)

In the case of amplifier operation, this knowledge is necessary for the following reasons: to ensure operation with $V_{CB} > 0$ V; to control P_C; and to control the parameters of the incremental model (*see* next chapter).

Satisfactory biasing takes into account uncertainties in β_0 and V_{BE} and variations in these parameters through changes in environmental conditions, notably temperature. With discrete BJT circuits, three possible biasing techniques are: fixed $-V_{BE}$, single base-resistor; and base-potentiometer. Another technique, only possible with ICs, is considered after these have been discussed.

Fixed-V_{BE} biasing (*see* Fig. 3.12)

Figure 3.13 shows the transfer characteristics of two samples X, Y of identically-coded BJTs that may be used in this circuit. X and Y have base characteristic current parameters I_{SX}, I_{SY}, respectively, and their V_{BE}s at a specified collector current $I_C = I_P$, are V_X and V_Y. Hence,

$$I_P = I_{SX}e^{V_X/V_T} = I_{SY}e^{V_Y/V_T} \tag{3.19}$$

If the BJTs are operated at $V_{BE} = V_{BB}$, their collector currents I_{CX}, I_{CY} are given by,

$$I_{CX} = I_{SX}e^{V_{BB}/V_T} \tag{3.20}$$

$$I_{CY} = I_{SY}e^{V_{BB}/V_T} \tag{3.21}$$

It follows from Equations (3.19), (3.20) and (3.21) that,

$$I_{CX}/I_{CY} = I_{SX}/I_{SY} = e^{\Delta V_{BE}/V_T} \tag{3.22}$$

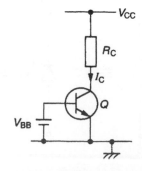

Fig. 3.12   Not recommended: fixed-V_{BE} biasing

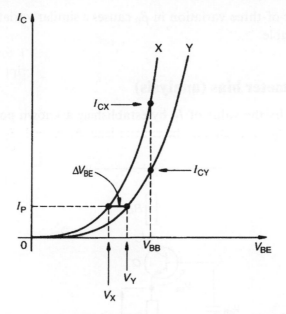

Fig. 3.13 Showing that a small spread in V_{BE} can cause a large spread in I_C

where,

$\Delta V_{BE} = (V_Y - V_X) = $ 'V_{BE}-match'

If $\Delta V_{BE} = 30$ mV then,

$(I_{CX}/I_{CY}) > 3$ (3.23)

This collector current spread is unacceptable. Furthermore, even if $\Delta V_{BE} = 0$ the scheme is still unsuitable because I_S is not known to the circuit designer and is very temperature sensitive.

Single base-resistor (Fig. 3.14)

$I_C = \beta_0 I_B = \beta_0 (V_{CC} - V_{BE})/R_B$ (3.24)

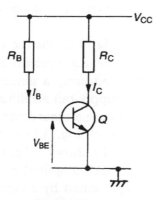

Fig. 3.14 Not recommended: single base-resistor biasing

A possible factor-of-three variation in β_0 causes a similar variation in I_C, so this method is unsuitable.

Base-potentiometer bias (analysis)

This attempts to fix the value of I_E by establishing a known potential difference across a resistor R_E connected in the emitter lead. As $\alpha_0 \simeq 1$, I_C is fixed also.

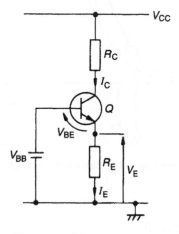

Fig. 3.15 V_{BB} and R_E fix emitter current I_E

The origin of the scheme is the two-battery arrangement of Fig. 3.15. By inspection,

$$V_E = (V_{BB} - V_{BE})$$

Hence,

$$I_C = \alpha_0 I_E = \alpha_0 (V_{BB} - V_{BE})/R_E \qquad (3.25)$$

It follows that $a \pm 50$ mV spread in V_{BE} causes an error in I_C of less than $\pm 5\%$ if $V_E > 20(\Delta V_{BE})$, i.e. $V_E > 1$ V. I_C is then effectively circuit-determined rather than device-determined.

In implementing the circuit of Fig. 3.15 the use of power supplies of both polarity offers significant benefits (*see*, e.g. Problem 5). For a single supply rail, V_{BB} in Fig. 3.15 can be implemented using a suitably biased zener diode (*see* Problem 7). However, in a single-supply-rail amplifier design it is not generally possible, for operational reasons, to use a zener diode. The circuit of Fig. 3.16 is then appropriate. The battery V_{BB} of Fig. 3.15 is replaced by a base-potentiometer network R_1, R_2. Figure 3.17 shows a d.c. equivalent circuit used for analysis. It incorporates the EM model, simplified in that the base-emitter diode drop is taken as constant and represented by a battery. The circuit inside the dotted contour shows the Thévenin representation of the base bias network.

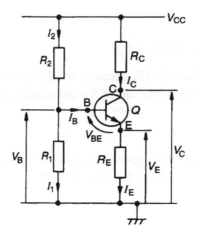

Fig. 3.16 Base-potentiometer biasing scheme

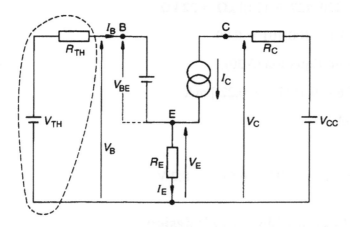

Fig. 3.17 D.C. equivalent circuit of Fig. 3.16

Thus,

$$V_{TH} = R_1 V_{CC}/(R_1 + R_2) \tag{3.26}$$

and,

$$R_{TH} = R_1 R_2/(R_1 + R_2) \tag{3.27}$$

Applying Kirchhoff's Voltage Law to the base circuit,

$$V_{TH} = I_B R_{TH} + V_{BE} + V_E \tag{3.28}$$

Substituting $I_B = I_E/(1+\beta_0) \simeq I_E/\beta_0$, $V_E = I_E R_E$ gives,

$$I_E[R_E + (R_{TH}/\beta_0)] = (V_{TH} - V_{BE}) \tag{3.29}$$

But,

$$I_E = I_C/\alpha_0$$

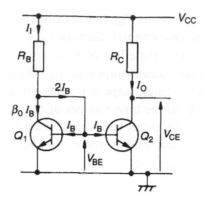

Fig. 3.18   A basic 'current-mirror' circuit

Biasing BJTs inside ICs

Figure 3.18 shows the simplest bias scheme of the 'current-mirror' family, which is popular in analogue IC design. The output current I_O, 'mirrors', closely, the input current, I_1, because of the excellent matching in the respective terminal characteristics of the two BJTs, Q_1, Q_2 fabricated in close proximity on the same semiconductor chip. Q_1, Q_2 have the same emitter area, A_J, and operate with the same base–emitter voltage, V_{BE}. Hence, from Equation (3.10), they pass the same base current, I_B. At the collector of Q_1, KCL gives,

$$I_1 = \beta_0 I_B + 2I_B = (\beta_0 + 2)I_B \tag{3.31}$$

Ignoring the Early voltage of Q_2,

$$I_O = \beta_0 I_B \tag{3.32}$$

Thus,

$$\lambda = (I_O/I_1) = \beta_0/(\beta_0 + 2) \simeq [1 - (2/\beta_0)] \tag{3.33}$$

i.e. $\lambda \simeq 1$.

If the Early voltage, V_A, of Q_2, is allowed for,

$$I_O = \beta_0 I_B[1 + (V_{CE}/V_A)] \tag{3.34}$$

Then, from Equations (3.31) and (3.34),

$$\lambda = \beta_0[1 + (V_{CE}/V_A)]/(\beta_0 + 2) \tag{3.35}$$

Using the binomial expansion and ignoring second-order terms, Equation (3.35) reduces to,

$$\lambda \simeq [1 - (2/\beta_0) + (V_{CE}/V_A)] \tag{3.36}$$

A sketch of I_O vs V_{CE} is shown in Fig. 3.19 for the useful output range $BV_{CEO} > V_{CE} \geq 0.7 \text{ V}$.

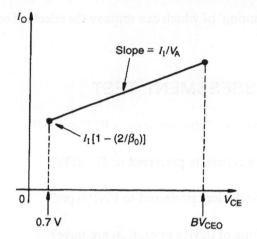

Fig. 3.19   Output characteristics for Fig. 3.18

If, in Fig. 3.18, the emitter area of Q_2 is larger than that of Q_1 by a factor m ($\ll \beta_0$), then it can be shown (*see* Problem 13) that,

$$\lambda \simeq m[1 - \{(1 + m)/\beta_0\} + (V_{CE}/V_A)] \tag{3.37}$$

The magnification factor of the 'mirror' is approximately m. The current-mirror finds wide practical application, e.g. in voltage-to-current converters, linear-sawtooth voltage waveform generation, temperature-transducer design. [Cautionary note: the current-mirror cannot be constructed using discrete transistors because there is no guarantee of acceptable V_{BE}-matching. There are, however, some small scale ICs such as the CA3046 that offer an array of several transistors that can be used effectively in such applications.]

Example 3.3

Design a current-mirror, based on the circuit of Fig. 18, to give $I_O = 0.1$ mA. (Assume: $\beta_0 = 100$; $V_{BE} = 0.65$ V; $V_{CC} = 10$ V; $V_A = \infty$.)

Solution

$I_O = 0.1$ mA, $I_B = 0.1$ mA/100. Hence,

$I_1 = [0.1 + (2 \times 0.001)]$ mA $= 0.102$ mA.

$\therefore R_B = [(10 - 0.65)$ V/0.102 mA$] = 91.67$ kΩ

Thus,

$R_B = (91$ k$\Omega + 680$ $\Omega)$

General note: in some instrumentation applications (*see*, e.g. Problem 7) it is necessary to set a current to a precise value. This entails the use of a potentiometer, the 'trimming' of which can remove the effects of component and supply rail tolerances.

3.7   SELF ASSESSMENT TEST

1 What is the meaning of the arrow on a BJT symbol?

2 Why are Si BJTs generally preferred to Ge BJTs?

3 Why are NPN Si devices preferred to PNP types?

4 How many regions of device operation are there?

5 What is the boundary line between the saturation region and the forward-active region?

6 What is meant by the description 'mutual characteristic'?

7 If $\alpha_0 = 0.98$ what is β_0?

8 If $\beta_0 = 200$ what is α_0?

9 What is meant by 'Early voltage'?

3.8   PROBLEMS

(Assume the following NPN transistor data apply in numerical questions, unless otherwise stated: $\beta_0 = 100$; $V_{BE} = 0.65$ V; $\Delta V_{BE} = \pm 50$ mV, where applicable; $V_A = \infty$).

1 Devise a chart similar to that of Table 3.1 but with entries expressed in terms of α_0 rather than β_0.

2 Show, using Equations (3.9), (3.10) and (3.11) that,

$$\alpha = (I_C/I_E) = \alpha_0[1 + (V_{CB}/V_A)]/[1 + (\alpha_0 V_{CB}/V_A)]$$

Hence, derive Equation (3.17) of the text. (Hint: use the binomial expansion, ignore second and higher order terms and assume $\beta_0 \gg 1$.)

3 Derive Equation (3.18).

4 Calculate $P_C(\text{max})$ for a BC107 at $T_A = 50°C$.

5 Calculate I_C, V_C for the circuit of Fig. 3.20. Determine also the maximum and minimum values of I_C, taking into account ΔV_{BE}, a $\pm 1\%$ tolerance on R_E, and $400 \geqslant \beta_0 \geqslant 100$.

6 Calculate I_{C1}, for the circuit of Fig. 3.21, assuming Q_1, Q_2 have identical characteristics.

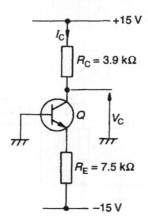

Figure 3.20

7 Figure 3.22 shows an ohmmeter. A current generator provides an output current I_C that is used to produce a potential difference, V_X, across a resistor R_X

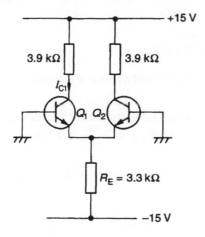

Figure 3.21

whose value is to be determined. The digital voltmeter, DVM, which monitors V_X, is required to read 1.000 V when $R_X = 10$ kΩ.

Calculate suitable values for R_E, R_V so that this setting can be achieved. Calculate, also, the maximum value of R_X that can be indicated. (Assume the NPN device data, given above, applies also to Q_2, Q_3. For Q_1, assume $BV_{BR} = 6$ V ± 0.25 V. Neglect loading effects of the DVM.)

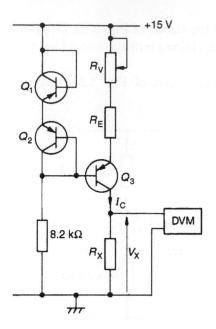

Figure 3.22

8 For the circuit of Fig. 3.16 assume: $V_{CC} = 15$ V; $R_E = 2.2$ kΩ; $R_C = 2.7$ kΩ. Calculate I_C, V_C for,

    (a) $R_1 = 1.5$ kΩ, $R_2 = 3.3$ kΩ
    (b) $R_1 = 15$ kΩ, $R_2 = 33$ kΩ
    (c) $R_1 = 150$ kΩ, $R_2 = 330$ kΩ

Comment on the discrepancies between the results.

9 Using Table 3.3, design a base-potentiometer bias circuit to give, $I_C \simeq 2.5$ mA, $V_C = 8$ V. (Assume $V_{CC} = 15$ V, 2 V $\geqslant V_E \geqslant 1$ V)

10 Show, for the circuit of Fig. 3.23, that I_C and V_C are given by the solution of the following simultaneous equations,

$$V_C[(1/R_C) + (1/R_2)] + I_C = (V_{CC}/R_C) + (V_{BE}/R_2)$$

$$(V_C/R_2) - (I_C/\beta_0) = V_{BE}[(1/R_1) + (1/R_2)] + (V_{BB}/R_1)$$

(Hint: apply KCL at the collector and base of Q.)

11 If, in 3.10, it is assumed that $\beta_0 = \infty$, show that,

$$I_2 = I_1 = (V_{BB} + V_{BE})/R_1$$

$$V_C = (V_{BB}R_2/R_1) + [(R_1 + R_2)/R_1]V_{BE}$$

$$I_R = (V_{CC} - V_C)/R_C$$

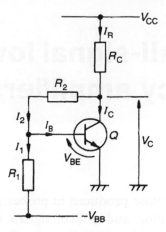

Figure 3.23

$$I_C = I_R - I_2$$

Suggest a circuit condition that justifies the approximation $\beta_0 = \infty$.

12 Figure 3.24 shows a current generator circuit, employing a current-mirror, giving an output current that is insensitive to possible variations in V_{CC}. Q_1, Q_2 and Q_3 may be assumed to have identical characteristics. Calculate I_C for $V_{XX} = 5$ V and for $V_{XX} = 10$ V.
   (Assume $V_Z = 5.6$ V, $V_A = 100$ V)

13 Derive Equation (3.37).

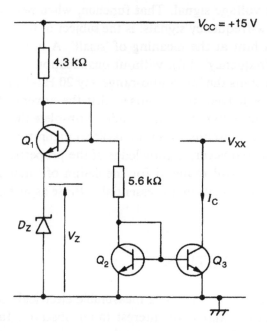

Figure 3.24

4 BJT small-signal low-frequency amplifiers

Microvolt signals, such as those produced in probes placed on a human scalp during neurological examination, and millivolt signals, like those produced by the action of a microphone, are magnified for recording purposes by electronic amplifiers. Amplifier analogies in other branches of engineering science include the mechanical lever, used to magnify small displacements, and the microscope, which gives optical enlargement of small objects.

There are numerous types of electronic amplifier (just called 'amplifier' from now on) and ways of classifying them, e.g. according to frequency range and to power-handling capability. A number of these are considered as the subject unfolds.

By definition, all amplifiers must be able to exhibit power-gain and this entails the use of active devices, namely, transistors. However, so called 'power-amplifiers' are mainly required at the outputs of systems, to drive a loudspeaker or an electric motor, etc. More often there is a need for a 'voltage amplifier', whose principal function is to produce an output voltage signal that is a magnified replica of an input voltage signal. That function, when performed by a BJT on small-amplitude low-frequency signals, is the subject of the present study. There has already been a hint at the meaning of 'small'. A more precise definition is given later. 'Low-frequency' (l.f.), without qualification, is also vague. For the present, assume it means the low audio-range, say 20 Hz–1 kHz.

It must be acknowledged at the outset, that the advent of ICs, such as the Operational Amplifier (Chapter 9), has rendered obsolete the design of a range of discrete low-frequency circuits using the basic configurations that are described later in this chapter. However, a knowledge of the properties and limitations of these configurations is still required for the design of 'interface' circuits at the inputs and outputs of systems and is essential to those aspiring to be designers of analogue ICs.

4.1 AMPLIFIER ACTION

In this book, a 'signal' is defined as a measurable electrical quantity that has an arbitrary time-variation and is of interest to an observer. In this context it is synonymous with 'waveform'.

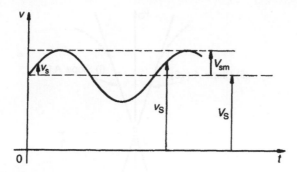

Fig. 4.1 Signal notation

The particular signal shown in Fig. 4.1 is useful in the classification of voltage amplifiers. It is labelled in accordance with the standard notation given at the beginning of this book. Thus,

v_S = total instantaneous value = $(V_S + v_s)$.
V_S = time-independent, i.e. constant or d.c. component of v_S
v_s = instantaneous value of time-varying component
V_{sm} = peak value of v_s

If, as shown, the time varying component v_s is sinusoidal then its RMS value is written V_s.

When it is required to amplify the total instantaneous value of an input signal that contains a non-zero d.c. component a d.c. amplifier is used.

The output voltage, v_O, is related to the input voltage, v_I, by a numerical factor that is known as the voltage gain and given the symbol A (amplification) or G (gain).

$$A = v_O/v_I \tag{4.1a}$$

or,

$$(V_O + v_o) = A(V_I + v_i) \tag{4.1b}$$

The d.c., or zero-frequency component, of the input signal is thus amplified by the same amount as any frequency component up to some designable upper limit f_H. The full-line graphs in Fig. 4.2 show ideal transfer characteristics of two amplifiers: graph (i) is for the case $A > 0$; graph (ii) for the case $A < 0$. The dotted lines show the departure from linearity that occurs in practical amplifiers for large input signals. Figure 4.3 shows the variation in gain magnitude with frequency (the 'frequency response'): discussion of this and related frequency plots is deferred till Chapter 7. Returning to Fig. 4.2, the negative sign for 'A' means that the output is inverted with respect to the input *not* that there is a reduction in signal amplitude in passing through the amplifier. Thus, the output for graph (ii) is amplified by a factor $|A|$ and turned upside down with respect to the input. In the case of signal variations that are sinusoidal this means a phase

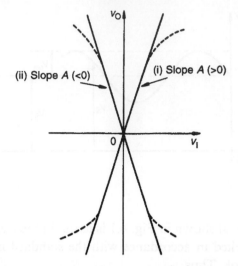

Fig. 4.2 Voltage amplifier transfer characteristics. Dotted curves show practical departure from linearity

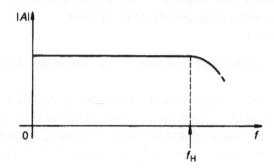

Fig. 4.3 Frequency response for the gain of a d.c. amplifier

shift of π radians (but *not* a time shift). The inverting action is well-illustrated by a general 'cross-plotting' procedure that is used for displaying the detail of an output signal.

Figure 4.4(a) is a repeat of graph (ii) of Fig. 4.2. In Fig. 4.4(b) the input is plotted to the same voltage scale as that in Fig. 4.4(a). The time axis is that for v_O extended vertically downwards, from Fig. 4.4(a), and re-scaled. Similar comments apply to the voltage and time scales in Fig. 4.4(c). Any chosen point P_X of the input signal cross-plots to a point P'_X of the output signal: hence, the output signal can be constructed from the input signal. This plotting procedure makes clear not only signal inversion but also the distortion that can arise, with large signals, due to any non-linearity in the transfer characteristic.

On system diagrams, amplifiers are generally represented by triangular or box symbols (*see* Fig. 4.5(a)) with the letter 'A' inscribed. Unless otherwise indicated, the direction of signal flow is conventionally taken as being from left to right. An equivalent circuit for a d.c. amplifier is shown in Fig. 4.5(b). R_I, R_O are the input

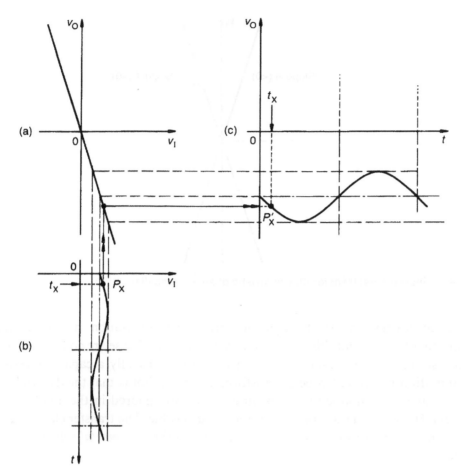

Fig. 4.4 Showing how an output signal, (c), can be derived from the transfer characteristic, (a), and input signal, (b), by 'cross-plotting'

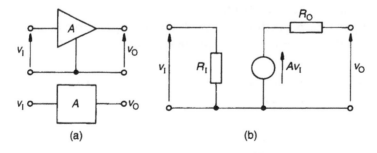

Fig. 4.5 Amplifier representation: (a) block schematic; (b) circuit equivalent

and output resistances, respectively: for an 'ideal' voltage amplifier, $R_I = \infty$, $R_O = 0$.

D.C. amplifiers are essential in some applications, e.g. to amplify the signals obtained from thermocouples, but they present problems in design and use. A notable problem is a horizontal shift in the position of the transfer characteristic,

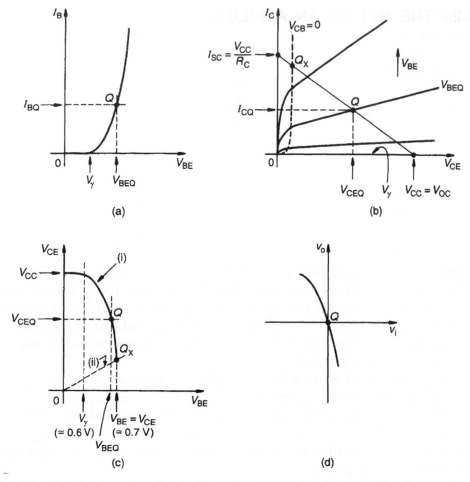

Fig. 4.9   Showing how the circuit of Fig. 4.8 can provide incremental voltage gain at quiescent point Q. (a) Input characteristic. (b) Output characteristic. (c) D.C. transfer characteristic. (d) Incremental transfer characteristic

voltage gain must exist in that region. The d.c. transfer characteristic is not linear for the complete range $0.7\,\text{V} > V_{BE} > 0.6\,\text{V}$. However, it can be considered *incrementally linear* for changes in V_{BE} that are small compared with $100\,\text{mV}$, about a bias point such as Q. The BJT is usually biased so that $V_{CEQ} \simeq (V_{CC} - V_{EQ})/2$. This condition ensures the availability of the maximum output voltage change (peak-to-peak 'swing') for sinusoidal variations.

$V_{EQ} = 0$ in the simplified circuit of Fig. 4.8. In the general case of base-potentiometer biasing, it is desirable to have $V_{EQ} \geq 1\,\text{V}$ for good definition of I_{CQ} (*see* Chapter 3). However, V_{EQ} should not exceed about 20% of V_{CC} because of the consequent reduction in available output swing. On the incremental transfer characteristic (Fig. 4.9(d), derived from Fig. 4.9(c)) Q is the zero point for v_i, v_o.

Graphical techniques using load lines are instructive in showing how voltage gain is achieved, but they are not very often used in the design of l.f. voltage amplifiers using BJTs. Apart from the tedium and inaccuracies of plotting, there are three fundamental reasons why this is so. First, a given set of characteristics varies widely from one device to another of a given type. Second, theoretical limits to incremental gain are not obvious from graphical plots. Third, device $I-V$ curves are only applicable to d.c. and low frequencies.

Amplifier analysis and design is based on the use of incremental equivalent circuits obtained from a knowledge of device operation that is common to all BJTs.

4.3 INCREMENTAL EQUIVALENT CIRCUITS

In this section the concept of an incremental equivalent circuit, touched on obliquely in Chapters 1 and 2, is put on a firmer footing using the PN junction diode as an example.

The CE incremental circuit for the BJT is derived from the EME model. Finally rules for drawing an incremental equivalent circuit for a BJT amplifier stage are formulated so that incremental performance parameters (e.g. voltage gain) can be calculated.

Diode incremental model

For the series circuit of Fig. 4.10(a) the diode quiescent current I_Q, and voltage, V_Q, are related by

$$V_{AA} = I_Q R + V_Q \tag{4.3}$$

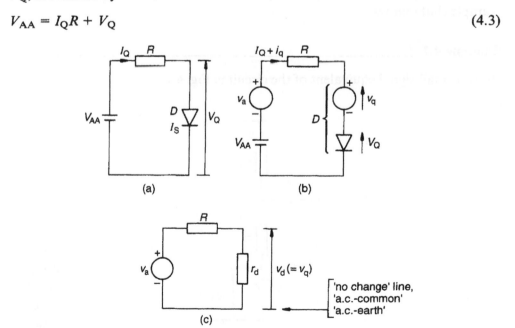

Fig. 4.10   Derivation of incremental model of a diode circuit. (a) D.C. conditions. (b) Total instantaneous equivalent circuit. (c) Incremental equivalent circuit

Suppose a change v_a in V_{AA} causes changes i_q, v_q in I_Q, V_Q, respectively. Then,

$$(V_{AA} + v_a) = (I_Q + i_q)R + (V_Q + v_q) \tag{4.4}$$

This instantaneous relationship is illustrated in Fig. 4.10(b).

Subtracting Equation (4.3) from Equation (4.4) gives,

$$v_a = i_q R + v_q \tag{4.5}$$

From Section 1.1, $r_d = (V_T/I_Q)$ and $v_q = i_d r_d$ if $v_q \leq 5\,\text{mV}$. Taking this restriction, which defines 'small-signal', into account,

$$v_a = i_q(R + r_d) \tag{4.6}$$

Thus, i_q and v_q can be calculated provided v_a does not cause v_q to exceed $5\,\text{mV}$. The incremental equivalent circuit is shown in Fig. 4.10(c). Compared with the d.c. bias circuit of Fig. 4.10(a), there is no change in the respective positions of the components. However, V_{AA} is now replaced by its *change* v_a, and D is replaced by its incremental resistance r_d. R does not alter since its incremental and d.c. resistance are the same.

The 'no change' line, shown, is usually referred to as 'a.c.-common' or 'a.c.-earth' (also 'signal-earth', though this implies that 'signal' is only the time-varying part of an input). In this case it is also the d.c.-common line.

On a small-signal equivalent circuit *any* point at a fixed potential is regarded as being connected to the a.c.-common line because a change in current flowing to, or from, the point does not produce a change in voltage. This is shown in the example that follows.

Example 4.1 _____

Draw a small-signal equivalent of the circuit in Fig. 4.11.

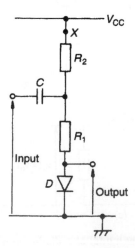

Fig. 4.11   Circuit for Example 4.1

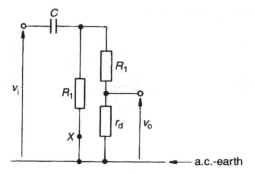

Fig. 4.12 Incremental model for Fig. 4.11

Solution

D is replaced by its incremental resistance r_d.

There is no change at point X in Fig. 4.11 when an input signal is applied so the upper end of R_2 is connected to a.c.-common as shown in the equivalent circuit of Fig. 4.12.

The use of incremental electrical parameters, rather than absolute values, has its counterparts in everyday life. One example concerns the lifting of a weight from a floor to the surface of a nearby table. The energy expended in doing this is dependent on the height of the table surface above the floor, not the height of the floor above sea level.

Incremental equivalent circuit of the BJT

Figure 4.13 is a pictorial summary of the way that a low frequency incremental model of the BJT can be derived from a d.c. model via a consideration of the slopes of the characteristics at an operating point and the interpretation of these slopes as circuit elements. Figure 4.13(a) shows the EME d.c. model. The characteristics in Fig. 4.13(b) are derived from Fig. 3.6 row (a). The mutual characteristics are shown here plotted to linear scales. The quiescent point is labelled Q throughout, and this letter is used as an additional subscript for bias quantities. Assuming the changes i_b, v_{be}, i_c and v_{ce} are small we can, as with the diode, approximate the curves by tangents at the operating point.

At the input to the BJT,

$$i_b = v_{be}/r_\pi \qquad (4.7)$$

where, r_π = incremental input resistance or,

$$r_\pi = 1/(\text{slope of } I_B, V_{BE} \text{ curve at } Q) \qquad (4.8)$$

At the output of the BJT, the *total* change, i_c, can be considered to be the sum of two components, i_1 and i_2. To represent these, only one set of curves is really necessary, the mutual characteristics or the output characteristics. Both sets are

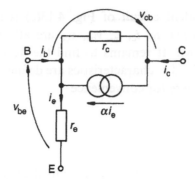

Fig. 4.15   Further, approximate, version of Fig. 4.13(c)

Fig. 3.6, row (b). It is obtained here by making use of Equation (4.16), from which we obtain $g_m = \beta/r_\pi$. Hence,

$$g_m v_{be} = (\beta/r\pi)v_{be} = \beta i_b$$

(Note: β, rather than β_o.)

Another equivalent circuit, appropriate to the CB characteristics of Fig. 3.6, row (c), is shown in Fig. 4.15. The incremental output circuit can be derived from Equation (3.17) (*see* Problem 3). For $\beta_0 V_A \gg V_{CB}$ the incremental output resistance, r_c is given by

$$r_c = \beta_o V_A / I_{CQ} \tag{4.19}$$

The incremental input circuit can be found from Equation (3.16) by ignoring the small dependence of I_E on V_{CB}.

Since I_E is exponentially dependent on V_{BE}, the incremental input resistance, r_e, seen at the emitter terminal is,

$$r_e = V_T/I_{EQ} = \alpha/g_m \tag{4.20}$$

To complete the l.f. equivalent circuits of the BJT it is necessary to add, to each of them, the extrinsic base resistance, r_x, mentioned in the last chapter. Thus, a modified form of Fig. 4.13(c) is shown in Fig. 4.16. The junction point of r_x and r_π

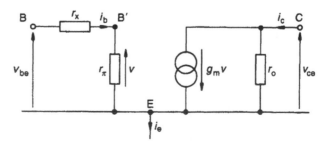

Fig. 4.16   The addition of extrinsic base resistance, r_x, completes the BJT l.f. model

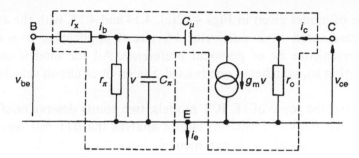

Fig. 4.17   'Hybrid-π' equivalent circuit, used in high frequency work

is usually labelled B', but to avoid the use of a primed subscript it is convenient to write v_π (or, as here, v with no subscript) instead of $v_{b'e}$.

The reader may wonder why it is necessary to include r_x in the model as it is neglected in d.c. calculations. The reason is as follows. At $I_{CQ} = 5\,mA$, $g_m = 200\,mS$. Taking $\beta_0 = 100$, Equation (4.16) gives $r_\pi = 500\,\Omega$: r_x, typically $100\,\Omega$, is not negligible compared with this, and the situation is worse at higher current levels. On the other hand, the d.c. drop across r_x is only some $5\,mV$ which is negligible compared with V_{BE} ($\sim 0.7\,V$).

A high frequency equivalent circuit of the BJT is given in Fig. 4.17. It is Fig. 4.16 with the addition of C_π, C_μ ($\ll C_\pi$) which take into account, respectively, capacitance effects associated with the base–emitter junction and collector–base junction. C_π and C_μ can be deduced from data sheet information (*see* Chapter 7). As far as the BJT is concerned, 'low-frequency' means that range over which the presence of C_π, C_μ can be ignored because the currents in them are negligible in comparison with the currents in the other components of the model.

The circuit of Fig. 4.17, and by implication the l.f. version of Fig. 4.16, is known as the *hybrid-π* model of the BJT. The dotted outline indicates why the Greek letter π is used in the description. 'Hybrid' means 'mixed' and the parameters used in the model have mixed dimensions. Thus, r_x, r_π, r_o are expressed in ohms and g_m in Siemens (mS, in low power circuits).

The hybrid-π circuit is extensively used in BJT circuit analysis and design because its parameters are related to device geometry and doping levels and have well-known dependencies on I_{CQ}, V_{CEQ} and T. Furthermore, the parameters are frequency-independent up to several hundred MHz. The I_{CQ} dependence emphasizes the necessity of 'good' biasing if parameters are to be deduced from d.c. operating conditions. In this context, g_m is particularly important. From Equation (4.15), all BJTs operating at the same I_{CQ}, T, have the same g_m. At room temperature,

$$g_m\,(mS) \simeq 40 I_C\,(mA) \tag{4.21}$$

Thus, at $I_{CQ} = 2\,mA$, $g_m \simeq 80\,mS$.

The choice of model given in Figs 4.13(c), 4.14 and 4.15, with the addition of r_x in each case, used in a particular l.f. circuit application is a matter of analytical convenience or of personal preference. All the models contain the same information so, whichever one is used, the result of circuit calculations will be the same.

Before leaving the topic of l.f. BJT models, two points deserve brief mention. First, in the early days of transistor circuit analysis the BJT was regarded as a 'black box' characterized by 'h-parameters'. Since these offer less information than the hybrid-π parameters they are now rarely used. Second, with older types of BJT it was necessary to include a resistor r_μ in parallel with C_μ in the hybrid-π model, to take into account recombination of charge carriers in the base region. No significant error is made in neglecting this with modern low power transistors.

Example 4.2

Making reasonable assumptions, draw l.f. equivalent circuits, based on Figs 4.13(c), 4.14 and 4.15, for a BJT operated at room temperature with $I_{CQ} = 2\,\text{mA}$.

Solution

Assume $r_x = 100\,\Omega$, $V_A = 100\,\text{V}$, $\beta = 100$. Then,

$$g_m = (40 \times 2)\,\text{mS} = 80\,\text{mS: } r_\pi = (100/80 \times 10^{-3})\,\Omega = 1.25\,\text{k}\Omega$$

$$r_o = (100\,\text{V}/2\,\text{mA}) = 50\,\text{k}\Omega; r_c = (100 \times 50\,\text{k}\Omega) = 5\,\text{M}\Omega$$

$$\alpha = \beta/(1 + \beta) = 0.99; r_e = (25\,\text{mV}/2\,\text{mA}) = 12.5\,\Omega$$

The appropriate l.f. equivalent circuits are shown in Fig. 4.18.

Incremental model of an amplifier stage

The single stage amplifier circuit of Fig. 4.19(a) is used here to illustrate a general procedure for obtaining the incremental equivalent circuit from a circuit diagram. The purpose of each component is as follows. R_1, R_2 and R_E provide base-potentiometer biasing. R_C is the collector load resistor, across which output voltage variations are developed. C_B prevents the d.c. component of the input from upsetting the bias conditions but provides a low impedance path for coupling the input signal to the base of Q.

Since it is the signal voltage change between the base and emitter that is amplified it is important, for maximum gain, to minimize any voltage variation across R_E. This is achieved by the presence of a 'decoupling' capacitor C_E. If the reactance of C_E is sufficiently low, the voltage variation across it is negligible so the emitter of Q is effectively at a.c.-earth potential: C_E is sometimes called an emitter 'bypass' capacitor since emitter current variations

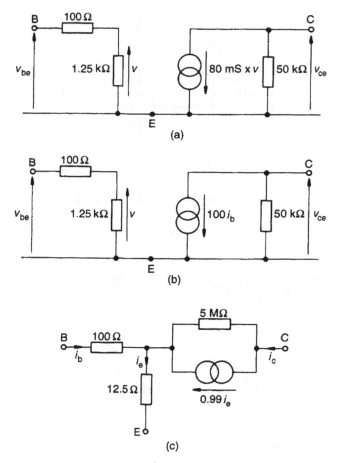

Fig. 4.18   Equivalent circuits relating to Example 4.2

bypass the emitter resistor R_E. The incremental equivalent circuit of Fig. 4.19(a) is shown in Fig. 4.19(b) and is constructed from the circuit diagram using the following general procedure, which is a development of the discussion in the previous section.

(a) Draw an l.f. small-signal equivalent circuit of the BJT.
(b) Identify the a.c.-earth line.
(c) Insert the components external to the BJT, working out from each terminal in turn and maintaining the same relative positions that the components have in the circuit diagram.
(d) Connect to a.c.-earth any point on the circuit diagram that is at a fixed d.c. potential.
(e) Redraw the circuit in a simplified form if the capacitor choice allows this.

Steps (a) to (d) have been followed in deriving Fig. 4.19(b): step (e) is discussed in the next section.

Let $R_{eq} = R_1 \| R_2$. Then, by the current divider rule,

$$I_b = I_i R_{eq}/(R_{eq} + r_x + r_\pi) \tag{4.22}$$

By the voltage divider rule,

$$V = V_i r_\pi/(r_x + r_\pi) \tag{4.23}$$

At the collector terminal,

$$I_o = I_c = -V_o/R_C \tag{4.24}$$

and

$$-(V_o/R_C) = g_m V + (V_o/r_o) \tag{4.25}$$

By algebraic manipulation of Equations (4.22) to (4.25), we can derive expressions for parameters such as R_i that define the performance of the stage. That procedure is necessary to assess limit performance and is applicable in numerical calculations (*see* Example 4.3). However, complete algebraic expressions tend to obscure the applicability of some simpler relationships that hold for the approximations usually made in practice. These simpler expressions hold for the following inequalities: $R_{eq} \gg (r_x + r_\pi); r_\pi \gg r_x; r_o \gg R_C$

$$R_i(\Omega) = (V_i/I_i) \simeq r_\pi = 25\beta/I_{CQ}(mA) \tag{4.26}$$

$$R_o(k\Omega) \simeq R_C(k\Omega) \tag{4.27}$$

$$A_v = (V_o/V_i) \simeq -g_m R_C = -40 I_{CQ}(mA) R_C(k\Omega) \tag{4.28}$$

$$A_i = (I_o/I_i) \simeq -g_m r_\pi = \beta \tag{4.29}$$

$$A_p = |(I_o V_o)/(I_i V_i)| = |A_i A_v| \simeq 40\beta I_{CQ}(mA) R_C(k\Omega) \tag{4.30}$$

Figure 4.21(b) is a system representation of the circuit of Fig. 4.19(a). In this V_s, R_s, comprise the equivalent circuit of an external source. If R_s is allowed for, the overall gain, (V_o/V_s), is that given by Equation (4.28) multiplied by an attenuation factor $R_i/(R_i + R_s)$.

Example 4.3

The following data (given in Example 3.1) are applicable to the circuit of Fig. 4.19(a): $V_{CC} = 15\,V$; $V_{BE} = 0.7\,V$; $R_1 = 27\,k\Omega$; $R_2 = 120\,k\Omega$; $R_E = 1.8\,k\Omega$; $R_C = 6.8\,k\Omega$; $\beta = 100$; $r_x = 100\,\Omega$; $V_A = 100\,V$. Calculate R_i, A_v, A_i and R_o assuming the reactances of C_B and C_E are negligible at the operating frequency.

Solution

From the solution to Example 3.1, $I_{CQ} = 1\,mA$. Hence, $g_m = 40\,mS$ and $r_\pi = \beta/g_m = (100/40 \times 10^{-3})\,\Omega = 2.5\,k\Omega$. The equivalent circuit is shown in Fig. 4.22

$$R_{eq} = 27\,k\Omega \| 120\,k\Omega \simeq 22\,k\Omega$$

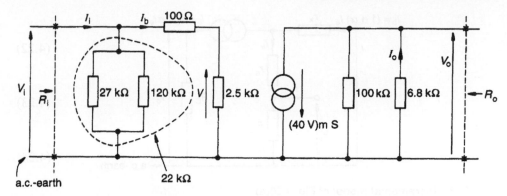

Fig. 4.22 Relating to Example 4.3

By inspection,

$R_i = 22\,\text{k}\Omega\|2.6\,\text{k}\Omega = 2.33\,\text{k}\Omega$

$R_o = 6.8\,\text{k}\Omega\|100\,\text{k}\Omega = 6.37\,\text{k}\Omega$

$I_i = V_i/(2.33\,\text{k}\Omega)$

$V = 2.5V_i/2.6 = 0.96V_i$

$g_m V = 40 \times 0.96V_i = 38.4V_i$

$V_o = -g_m V(100\,\text{k}\Omega\|6.8\,\text{k}\Omega) = -g_m V \times 6.37$

$\quad = -38.4 \times 6.37V_i = -257V_i$

$\therefore A_v = -257$

$\quad A_i = (I_o/I_i) = -(V_o/6.8\,\text{k}\Omega)/(V_i/2.33\,\text{k}\Omega)$

$\quad\quad = -(A_v \times 2.33)/6.8$

or,

$A_i = 88$

The CB configuration

A simplified equivalent circuit of the CB stage of Fig. 4.20(a) is shown in Fig. 4.23. It uses the CB circuit of Fig. 4.15 with r_c omitted. The capacitors are assumed to have zero reactance. We could equally well have chosen the CE equivalent circuits of Figs 4.13(c) and 4.14 with r_o omitted in both cases (*see* Problem 9).

The relevant circuit equations are:

$$V_i = -(I_e r_e + I_b r_x) \tag{4.31}$$

$$I_b = I_e(1 - \alpha) = I_e/(1 + \beta) \tag{4.32}$$

$$V_i = -I_e[r_e + \{r_x/(1 + \beta)\}] \tag{4.33}$$

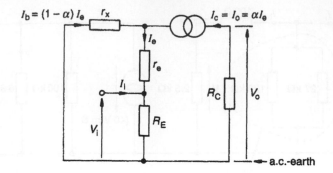

Fig. 4.23   Incremental model of Fig. 4.20(a)

and,

$$(V_i/-I_e) = r_e + \{r_x/(1 + \beta)\} \tag{4.34}$$

$$R_i = R_E||(V_i/-I_e) = R_E||[r_e + \{r_x/(1 + \beta)\}] \tag{4.35}$$

$$V_o = -I_o V_c = -\alpha I_e R_C \tag{4.36}$$

Equations (4.33) and (4.36) give: $A_v = (V_o/V_i)$; $A_i = (I_o/I_i) = -(V_o/R_C)/(V_i/R_i)$ $= -A_v(R_i/R_C)$; $A_p = |A_v A_i|$; $R_o = R_C$, by inspection.

The following simple expressions are valid for the practical approximations,

$$R_E \gg [r_e + \{r_x/(1 + \beta)\}], \; r_e \gg r_x/(1 + \beta)$$

$$R_i(\Omega) \simeq r_e = 25/I_{EQ}(mA) \tag{4.37}$$

$$R_o(k\Omega) \simeq R_C(k\Omega) \tag{4.38}$$

$$A_v \simeq \alpha(R_C/r_e) = +g_m R_C = 40 I_{CQ}(mA) R_C(k\Omega) \tag{4.39}$$

$$A_i \simeq -\alpha \simeq -0.99 \tag{4.40}$$

$$A_p \simeq 0.99 A_v \tag{4.41}$$

The negative sign for A_i means that an increase in I_i causes a decrease in I_o.

The system diagram shown in Fig. 4.21 is still applicable if the arrow indicating $|A_v| V_i$ is reversed.

If the base decoupling capacitor C_B in Fig. 4.20(a) is omitted, then the circuit equations are still valid provided R_{eq} ($= R_1||R_2$) is added to r_x wherever r_x appears. Note: generalizing Equation (4.34) we can say that, seen from the emitter, any resistance in the base circuit in series with r_x appears to be reduced by a factor $(1 + \beta)$.

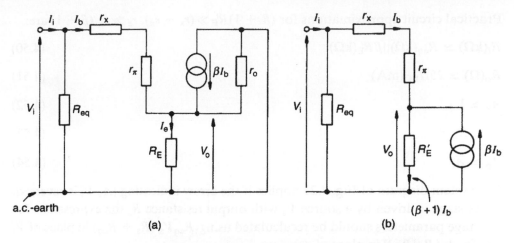

Fig. 4.24   Relating to Fig. 4.20(b): (a) incremental model; (b) simplified version of (a)

The CC configuration (the 'emitter-follower')

Figure 4.24(a) is an equivalent circuit of Fig. 4.20(b), assuming C_B has negligible reactance. Circuit operation is more easily appreciated from the simplified version shown in Fig. 4.24(b).

By inspection, the current in R'_E ($= R_E \| r_o$) is $(\beta + 1)I_b$

$$\therefore V_o = (\beta + 1)I_b R'_E \tag{4.42}$$

and,

$$V_i = I_b(r_x + r_\pi) + V_o = I_b(r_x + r_\pi) + (\beta + 1)R'_E \tag{4.43}$$

Hence,

$$V_i/I_b = (r_x + r_\pi) + (\beta + 1)R'_E \tag{4.44}$$

$$R_i = R_{eq}\|[(r_x + r_\pi) + (\beta + 1)R'_E] \tag{4.45}$$

and,

$$A_v = (V_o/V_i) = (\beta + 1)R'_E/[(r_x + r_\pi) + (\beta + 1)R'_E] \tag{4.46}$$

$$A_i = (I_e/I_i) = (V_o/R_E)(R_i/V_i) = A_v(R_i/R_E) \tag{4.47}$$

$$A_p = |A_iA_v| \tag{4.48}$$

If V_i is supplied by a source with zero output resistance, the output resistance of the CC stage, calculated in the same way as the input resistance of a CB stage, is,

$$R_o = R'_E\|[r_e + \{r_x/(1 + \beta)\}] \tag{4.49}$$

Practical circuit approximations for $(\beta + 1)R'_E \gg (r_x + r_\pi)$, $r_e \gg r_x/(\beta + 1)$ are:

$$R_i(k\Omega) \simeq R_{eq}(k\Omega)\|\beta R'_E(k\Omega) \tag{4.50}$$

$$R_o(\Omega) \simeq 25/I_{EQ}(mA) \tag{4.51}$$

$$A_v \simeq 1 \tag{4.52}$$

$$A_i \simeq R_i/R_E \tag{4.53}$$

$$A_p \simeq A_i \tag{4.54}$$

The system diagram of Fig. 4.21 applies if the arrow indicating $|A_v|V_i$ is reversed. If the stage is driven by a source V_s with output resistance R_s the expressions for the stage parameters should be recalculated using $R_{eq}V_s/(R_s + R_{eq})$ in place of V_i and $[r_x + (R_s\|R_{eq})]$ in place of r_x.

The name 'emitter-follower' arises because the output at the emitter follows the signal applied at the base ($A_v \simeq 1$). Note: generalizing Equation (4.44) we can say that, seen from the base, any resistance in the emitter circuit appears to be increased by a factor $(\beta + 1)$.

4.5 APPLICATIONS

The properties of the three basic configurations are compared in Table 4.1. The CE circuit, with a good mix of performance parameters, is the most widely used of the three arrangements. The CB stage offers good isolation between input and output. The CC stage provides a buffering action between a high resistance source and a low resistance load. A typical audio application is in the input circuit of a preamplifier driven by a ceramic/crystal microphone. Figure 4.25 shows a variation of the CE stage. The emitter resistor is split into two parts one of which, R_{EI}, is not bypassed by C_E. This enables an output to be taken from the emitter as well as the collector. By an extension of the analysis in the previous sections it can be shown that, under appropriate conditions, $(V_c/V_e) \simeq -R_C/R_{EI}$ (*see* Problem 11). This circuit, sometimes called a 'phase splitter', can be used to drive 'push-pull' output stages.

Table 4.1   Configurations compared

	CE	CB	CC
$|A_v|$	H	H	$\simeq 1$
$|A_i|$	H	$\simeq 1$	H
$|A_p|$	H	H	H
R_i	M	L	H
R_o	H	H	L

Key: H ≡ high; M ≡ medium; L ≡ low

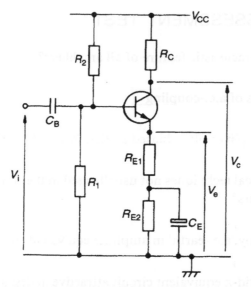

Fig. 4.25 CE stage with dual outputs

Figure 4.26(a) shows the collector circuit of a CE stage which has a base-potentiometer bias scheme and the collector coupled to a load R_X by a capacitor, C_X: in a two, or more, stage amplifier R_X represents the incremental input resistance of the next stage. If C_X is chosen so that $(1/\omega C_X) \ll R_X$ then the effective incremental load resistance seen at the collector is $R'_C = R_C \| R_X$ and, ignoring r_o, $A_v \simeq -g_m R'_C$. R'_C gives rise to an a.c. load line on the collector characteristics (*see* Fig. 4.26(b)). This passes through the quiescent point, Q, and has a slope $-(1/R'_C)$.

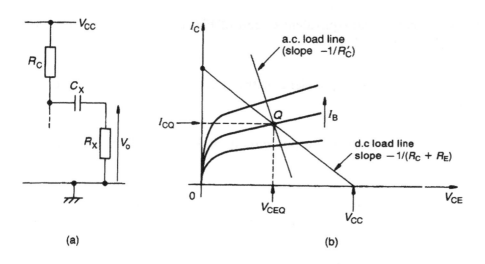

(a)

(b)

Fig. 4.26 CE stage with a.c.-coupled output: (a) circuit; (b) showing the a.c. load line

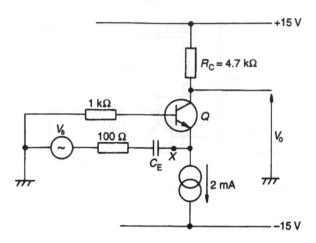

Figure 4.29

10 If, in Fig. 4.20(b), V_i is derived from a source V_s with internal resistance R_s, show that,

(a) $(V_o/V_s) = (\beta + 1)R'_E R_{eq}/[(r_x + R_x) + r_\pi + (\beta + 1)R'_E][R_{eq} + R_s]$

(b) $R_o = R'_E||[r_e + \{(r_x + R_x)/(\beta + 1)\}]$

where: $R'_E = R_E||r_o$; $R_{eq} = R_1||R_2$; $R_x = R_{eq}||R_s$.

11 Draw an incremental equivalent circuit of Fig. 4.25. Show that,

(a) $V_e \simeq V_i$

(b) $(V_c/V_e) \simeq -(R_C/R_{EI})$

Assume: $r_o = \infty$; $(\beta + 1)R_{E1} \gg (r_x + r_\pi)$; $\beta \gg 1$.

5   Field effect transistors (FETs): d.c.-characterization and biasing

The FET is the second type of semiconductor control device that has helped to catapult the subject of electronics from a subsidiary to a central role in engineering and scientific disciplines. It operates on an entirely different principle from that of the BJT, leading to some associated circuit properties, e.g. very high input impedance, than can be exploited by the equipment design engineer.

In a BJT the flow of *minority* carriers, across a base region from emitter to collector is dependent upon potential differences (V_{BE}, V_{CB}) that act in the direction of carrier flow. The FET, by contrast, is a semiconductor device in which the flow of *majority* carriers in a 'channel' between two terminals designated source (S) and drain (D) is dependent upon a *transverse* field resulting from the application of a potential difference between one of these terminals (normally, S) and a third, control, terminal known as the gate (G). A crude everyday analogy is a garden hose pipe, in which the flow of water can be controlled by a squeezing action in a direction perpendicular to the direction of flow. The particular way in which the electrical control function is achieved, without the requirement of significant d.c. input power, permits a division of FETs into two basic categories each of which can be further sub-divided.

This chapter considers the classification of FETs, discusses the characteristics of one popular type in some detail and shows how it can be biased for use in d.c. and in low-frequency small-signal amplifier applications.

5.1   THE FET FAMILY

The junction gate FET (Jugfet, JFET)

On each side of a PN junction there is a narrow layer devoid of charge carriers The P-side layer has a net negative charge resulting from fixed negatively-charged acceptor impurity atoms that have supplied holes which are elsewhere in the crystal lattice. Similarly, the N-side layer has an equal and opposite-polarity charge due to fixed positively-charged donor impurity atoms that have supplied electrons to the lattice.

This junction dipole-layer of charge is responsible for the depletion capacitance mentioned in Section 1.3. If the voltage across the junction changes then the widths of the depletion regions change, also, and it is the width-change that is made use of in the JFET. The voltage applied to a reverse-biased junction, associated with the gate terminal, alters the thickness of a conducting channel that exists between source and drain and, hence, the current flow for a given drain–source voltage. This channel may be of N-type or P-type material. The JFET is also known as a PN-FET.

The insulated gate FET (IGFET or MOSFET)

This type of device does not rely on a PN junction to achieve the control action. The voltage applied between the gate and source terminals determines the charge induced into the surface layer of a semiconductor material that is electrically isolated from the gate. This induced charge can establish, between source and drain, a conducting path where none existed before: this is the case with P-channel and N-channel 'enhancement-mode' devices.

Alternatively, the induced charge can alter the conduction characteristics of an already-existing (or 'built-in') channel: this is the case with P-channel and N-channel 'depletion-mode' devices.

From the start, IGFETs were fabricated using established Si bipolar processing technology and, as a result, had a metal gate electrode (aluminium) and used a thermally-grown silicon dioxide layer as an insulation material. This led to the two following device acronyms: MOST (metal oxide semiconductor transistor); MOSFET.

Although metal and oxide are not always used in manufacture, MOSFET is the most popular description of this type. CISFET (conductor, insulator, semiconductor FET) would, perhaps, be a better generic designation but, because of the extent of part usage in the technical literature, MOSFET is unlikely to be displaced.

FET family-relationships

The pear-shaped symbol shown in Fig. 5.1(a) represents an FET of unspecified type: the arrows show the sign convention for positive drain current (I_{DS}), drain–source voltage (V_{DS}), and gate–source voltage (V_{GS}). Figure 5.1(b) shows the general shape of the unified family of $|I_{DS}|$ vs $|V_{DS}|$ characteristics. In each case $|I_{DS}| \simeq 0$ when $V_{GS} = V_\gamma$: V_γ is thus a 'cut-in', 'cut-off' or 'theshold-of-conduction' voltage. The approximation sign is used because V_γ is often conveniently specified as that V_{GS} producing a stated low value of $|I_{DS}|$, e.g. a few μA.

For the JFET, V_γ is usually referred to as a 'pinch-off voltage' and written V_P. With the MOSFET, V_γ is usually referred to as the 'threshold voltage'. Although this is sometimes written V_T, it is better to use the terminology V_{TH} to avoid confusion with the thermal voltage (kT/q) introduced in Chapter 1.

When an FET is on and $|V_{DS}| < |V_{GS} - V_\gamma|$ it operates in the 'pre-pinch off',

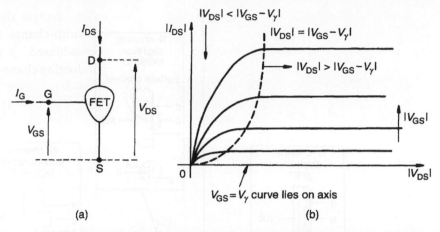

Fig. 5.1   (a) Symbolic representation; (b) output characteristics of an arbitrary type of FET

'triode', 'ohmic' or 'voltage-saturation' region: for $|V_{DS}| > |V_{GS} - V_\gamma|$, the FET operates in the 'pinch-off', 'constant-current', 'current-saturation' or 'pentode' region. This latter name stems from the similarity of the characteristics, in this region, with the output characteristics of a certain type of vacuum tube that was widely used many years ago.

The voltage-saturation region is used principally in switching applications, the constant-current region for linear applications.

The simplest relationship between I_{DS} and V_{GS} that gives a good first-order d.c. description of the FET in the constant-current region is that provided by the ideal 'square-law' model. Thus,

$$I_{DS} = K(V_{GS} - V_\gamma)^2 \qquad (5.1)$$

In Equation (5.1), K (with a subscript J or M to denote JFET and MOSFET, where appropriate) is a device parameter that is dependent upon material type, doping levels and geometry.

The polarities of the quantities in the relationship for the six possible types of FET are shown in Table 5.1. Equation (5.1) with the appropriate signs included is the basis of the FET classification chart of Fig. 5.2, which links the transfer, or

Table 5.1   FET parameter classification

$I_{DS} = K(V_{GS} - V_\gamma)^2$		Parameter polarity					Conditions
Device type	Carrier type*	K	I_{DS}	V_{DS}	V_{GS}	V_γ	
N-channel enhancement MOSFET	E	+	+	+	+	+	$V_{GS} > V_\gamma$
N-channel depletion MOSFET	E	+	+	+	+/−	−	$V_{DS} > (V_{GS} - V_\gamma)$
N-channel JFET	E	+	+	+	−	−	
P-channel JFET	H	−	−	−	+	+	$V_{GS} < V_\gamma$
P-channel depletion MOSFET	H	−	−	−	−/+	+	$V_{DS} < (V_{GS} - V_\gamma)$
P-channel enhancement MOSFET	H	−	−	−	−	−	

*H ≡ hole; E ≡ electron

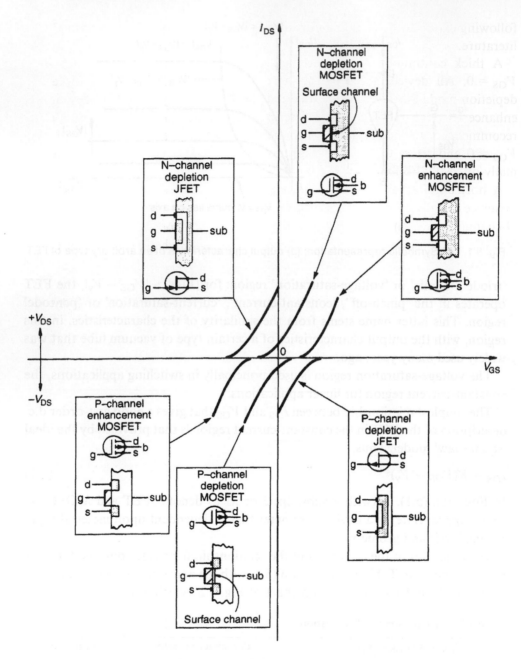

Fig. 5.2 **FET family relationships.** □, N semiconductor; ▓, P semiconductor; ▨, insulator; b, sub, 'substrate' or 'bulk'

mutual, characteristic with symbolic representation and simplified cross-section of device construction. In each case the channel region has the same polarity as the semiconductor material that comprises the source and drain. The symbols shown are unambiguous and permit instant recognition of device type without supplementary lettering or a word-description, as will be evident from the following explanation, so it is unfortunate that they are not always used in FET literature.

A thick continuous line connecting S and D indicates that $|I_{DS}| > 0$ for $V_{GS} = 0$. All devices so-drawn are termed 'depletion' FETs. In some texts depletion-mode MOSFETS are referred to as 'normally-on' and 'depletion/enhancement' devices but these descriptions can be confusing and are not recommended. A thick broken line connecting S and D means that $|I_{DS}| = 0$ for $V_{GS} = 0$ and refers to 'enhancement-mode' devices (otherwise known as 'normally-off' MOSFETS).

A horizontal arrow connected to a perpendicular line-section indicates the existence of a PN junction, the direction of the arrow being from P to N. Thus, an N-channel JFET has a P-type gate. In the case of a P-channel enhancement-mode MOSFET, the arrow points from the P-type induced channel to the N-type substrate.

The horizontal section of the gate terminal is always on the same level as the source. The MOSFET gate has a vertical section that is separated from the rest of the device symbol. This is to emphasize the electrical isolation of the terminal.

The substrate of the FET, shown as 'b' (for 'bulk' or 'body') on the symbol, is sometimes internally connected to the source and in such cases the substrate is often not shown. If a substrate is connected to a device terminal and is at the disposal of the circuit engineer, it must be connected to a potential which ensures that both the drain-substrate and source-substrate PN junctions do not become forward-biased.

For N-channel MOSFETs this potential is often, conveniently, the most negative in the circuit: for P-channel types the most positive potential in the circuit is appropriate.

General points concerning the way in which FETs complement the behaviour of BJTs, as active devices, are dealt with in the next chapter. MOSFETs are most often encountered in ICs, although discrete devices are used in radio receivers.

JFETs are widely used in discrete form, in analogue circuitry. For this reason the d.c. characteristics and biasing of the N-channel JFET are the main topics of the rest of this chapter: the treatment of P-channel devices is not given but closely mirrors this. Note: as the fabrication of all FETs is based upon technological processes developed for the production of Si BJTs and ICs, Ge FETs are not commercially available.

5.2 THE N-CHANNEL JFET: CONSTANT-CURRENT REGION

The ideal square-law model

Figure 5.3 contains redrawn parts of Figs 5.1 and 5.2 that are appropriate to an ideal N-channel JFET. In linear applications this is always operated with $V_{GS} \leq 0$

The drain characteristics for $V_{DS} > (V_{GS} - V_P)$ are approximated by their tangents at $V_{DS} = (V_{GS} - V_P)$. When extrapolated back they appear to originate from a common axis point $V_{DS} = -(1/\lambda):(1/\lambda)$, typically some tens of volts, is the FET equivalent of the Early voltage of BJT theory. The relevant equation for the output characteristics is,

$$I_{DS} = K_J(1 + \lambda V_{DS})[V_{GS} - V_P]^2 \tag{5.6a}$$

or,

$$I_{DS} = I_{DSS}(1 + \lambda V_{DS})[1 - (V_{GS}/V_P)]^2 \tag{5.6b}$$

where I_{DSS} is now the value of I_{DS} for $V_{GS} = 0$, extrapolated back to $V_{DS} = 0$. It is evident that $I_{DS} = 0$ when $(1 + \lambda V_{DS}) = 0$, for any V_{GS} (<0): furthermore, for a fixed V_{GS}, the plot of I_{DS} vs V_{DS} is a straight line.

The transfer characteristics for two arbitrary values of V_{DS}, namely, $V_{DS1} > (V_{GS} - V_P)$ and $V_{DS2} > (V_{GS} - V_P)$, are shown in Fig. 5.4(b). These can be derived from Fig. 5.4(a) by 'cross-plotting'.

It is still convenient to refer to the region $V_{DS} > (V_{GS} - V_P)$ as the constant-current region even though I_{DS} varies with V_{DS}.

Parameter temperature-dependence

In the constant-current region (at constant V_{GS}, V_{DS}) K_J, V_P, I_G all depend on temperature (T). From device physics,

$$(1/K_J)(dK_J/dT) = -m \tag{5.7}$$

$$(dV_P/dT) = -a(\simeq -2\,mV/°C) \tag{5.8}$$

$$(1/I_G)(dI_G/dT) = +b \tag{5.9}$$

In Equation (5.7) the letter m is used to emphasize the dependence of K_J on carrier mobility (i.e. the constant of proportionality relating carrier velocity to electric field strength).

The variations of V_P and I_G with T are given the same letter coefficients (a, b) as are used, respectively, for V_D and I_S in the case of the Si PN junction diode in Chapter 1 because they arise for similar physical reasons and have similar numerical values.

As indicated in Fig. 5.5, which shows the dependence of the transfer characteristic on T, there is an operating point Z (the 'compensation point') where the effects of T on V_P and K_J cancel and I_{DS} is temperature invariant. The co-ordinates of Z can be calculated from Equations (5.2), (5.7) and (5.8) (*see* Problem 4). However, the production spread in device characteristics means that it is rarely possible to exploit the existence of Z in circuit designs for mass production.

Although I_G (<0) is small at room temperature the exponential increase in $|I_G|$ with T can restrict the use of the JFET (as compared with the MOSFET) in some

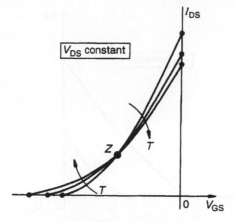

Fig. 5.5   Temperature-dependence of mutual characteristic

critical applications, e.g. electrometer measurements at elevated ambient temperatures.

5.3   PRACTICAL POINTS (N-CHANNEL JFETS)

Characteristics

It is not possible to determine a usable value of V_P by observing the point at which an experimental plot of I_{DS} vs V_{GS} 'intersects' the V_{GS} axis because the cut-off is not sharp.

A value of V_P most appropriate for use in Equations (5.2), (5.4) and (5.6) is obtained from a plot of $\sqrt{I_{DS}}$ vs V_{GS} extrapolated to $\sqrt{I_{DS}} = 0$, as shown in Fig. 5.6: a best straight line is drawn through points in the I_{DS} range of interest. The intercept on the vertical axis gives $\sqrt{I_{DSS}}$: from a knowledge of I_{DSS} and V_P, Equation (5.5) can be used to find an appropriate value for K_J.

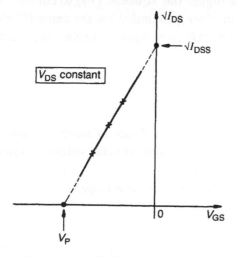

Fig. 5.6   Determination of V_P: crosses indicate measured data

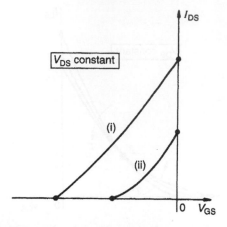

Fig. 5.7   Limit characteristics: (i) high current sample; (ii) low current sample

This measurement procedure, acceptable in a laboratory, is not easily auto-mated. If routinely carried out it would exact a cost premium on the tested unit, so device manufacturers normally specify a 'cut-off' voltage $V_{GS(OFF)}$ instead of V_P. $V_{GS(OFF)}$ is the easily-monitored value of V_{GS} required to reduce I_{DS} to some conveniently low test current, e.g. 1 nA.

As I_G is dependent on V_{DS}, as well as V_{GS} (<0), it is usually quoted for $V_{DS} = 0$ and a specified V_{GS} and given the designation I_{GSS}.

The d.c. characteristics and their defining parameters are subject to large production spreads (a $3:1$ ratio is not uncommon). The general effect on the mutual characteristics is shown in Fig. 5.7 for a given V_{DS}.

Curves (i) and (ii) represent, respectively, high and low current samples. As a memory-aid, imagine the transfer characteristic to be moved bodily along the V_{GS} axis.

Reverting to our hose pipe analogy, the larger the internal bore the larger the flow rate (I_{DSS}) but the bigger the 'squeeze' (V_P) to cut-off the flow. As with BJTs in ICs, JFETs made in close proximity on the same IC chip can exhibit close parameter-matching (e.g. $|\Delta V_{GS}| \simeq 5\,\text{mV}$ at a given I_{DS} and V_{DS}) and this is made use of in design work.

Ratings

The gate-channel PN junction is subject to zener breakdown and this must be avoided. The maximum gate-channel reverse voltage occurs at the drain end of the channel in amplifier operation.

The drain-source breakdown voltage BV_{DSX} at an arbitrary V_{GS} (<0) is given by,

$$BV_{DSX} = BV_{DSS} + V_{GS} \tag{5.10}$$

where, $BV_{DSS} = $ drain–source breakdown voltage for $V_{GS} = 0$.

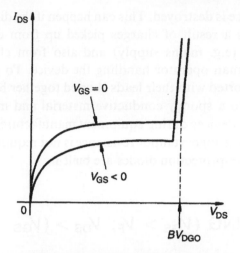

Fig. 5.8 Showing zener breakdown in output characteristics

BV_{DSS} is not always specified because its direct measurement could cause the dissipation of the FET to be exceeded. Fortunately, $BV_{DSS} \simeq BV_{DGO}$ (= drain–gate breakdown voltage with the source open circuit) and measurement of BV_{DGO} is relatively easy and does not cause excessive power dissipation. Hence, Equation (5.10) can be re-written,

$$BV_{DSX} \simeq BV_{DGO} + V_{GS} \tag{5.11}$$

Thus, for $BV_{DGO} = 25\,V$ and $V_{GS} = 0$, $BV_{DSX} = 25\,V$; for $V_{GS} = -1\,V$, $BV_{DSX} = 24\,V$. Equation (5.11) is illustrated in Fig. 5.8 for two values of V_{GS}.

The maximum reverse gate–source voltage permissible, BV_{GSS}, occurs at $V_{DS} = 0$: $BV_{GSS} \simeq BV_{DGO}$.

All the power dissipation, P_D, in amplifier use, occurs at the drain: $P_D = I_{DS}V_{DS}$. The power derating curve is similar in shape to that of a BJT.

Packaging

FETs, like BJTs, are produced in a range of packages, plastic, metal-can and stud-mounted, to suit power rating and intended market (e.g. domestic, industrial). They are coded as manufacturers' house-types and, where appropriate, given a '2N' designation.

Supplementary notes on the MOSFET

The characteristics and ratings of the MOSFET differ from those of the JFET principally in respect of I_G and gate-breakdown voltage. For the MOSFET, I_G is the leakage current of the oxide insulation: typically, $|I_G| \sim 1\,fA$ (i.e. $10^{-15}\,A$).

Gate-channel breakdown in a JFET is reversible if the power dissipation is kept within device limits: for a MOSFET, breakdown of the gate dielectric is nonreversible and the device is destroyed. This can happen if the dielectric breakdown voltage is exceeded as a result of charges picked up from external sources of electrical interference (e.g. mains supply) and also from charges accidentally transferred from a human operator handling the device. To guard against this, MOSFETS are transported with their leads shorted together by detachable metal springs or inserted into a spongy conductive material and manufacturers issue guidelines on handling devices during equipment manufacture.

Additionally, when the highest input resistance is not required, as in MOSFET logic circuits, shunt gate-protection diodes are built-in.

5.4   JFET BIASING ($V_{GS} > V_P$; $V_{DS} > (V_{GS} - V_P)$)

For amplifier and related applications, predictable bias values of I_{DS} and V_{DS} are necessary to ensure operation in the constant-current region, to control power dissipation and to fix the parameter values of the small-signal model (see next chapter).

The biasing technique must allow for wide variations in device characteristics and changes in ambient temperature. With discrete JFETs, three possible biasing techniques are: fixed-V_{GS}; automatic; gate-potentiometer. These are discussed and compared in the sections that follow, then a design procedure is presented.

Biasing circuits can be analysed graphically, if the mutual characteristics of an FET are provided. That method is straightforward but subject to drawing errors. A better method is to use free-hand sketches of the characteristics as a guide in an algebraic analysis using the ideal square-law FET model.

Fixed-V_{GS} bias

Figure 5.9 shows the connections and operating mechanism. In (a), V_X supplies the gate bias voltage. The circuit works with $R_G = 0$ but a non-zero R_G is included so that an a.c. voltage can be developed between gate and source in amplifier applications. (The capacitor, shown dotted, shows how the a.c. signal can be coupled to the gate terminal.)

R_D is the drain load resistor. A d.c. equivalent circuit is shown in (b). The potential difference across R_G due to I_G flowing in it is usually negligible compared with V_X, and will be ignored in the analyses. Thus if $I_G = -1\,nA$ and $R_G = 1\,M\Omega$, then $V_{GS} = -V_X + 1\,mV \simeq -V_X$.

The quiescent point Q is given, in (c), by the intersection of the gate–source load line $V_{GS} = -V_X$ with the mutual characteristic. The resulting drain current is I_{DSQ}. The quiescent drain-source voltage, V_{DSQ} in (d), is given by the intersection of the drain load line, whose equation is $I_{DS} = (V_{DD} - V_{DS})/R_S$, with the curve $V_{GS} = -V_X$. A necessary condition is $V_{DSQ} > (V_{GS} - V_P)$, i.e. $V_{DSQ} > -(V_X + V_P)$.

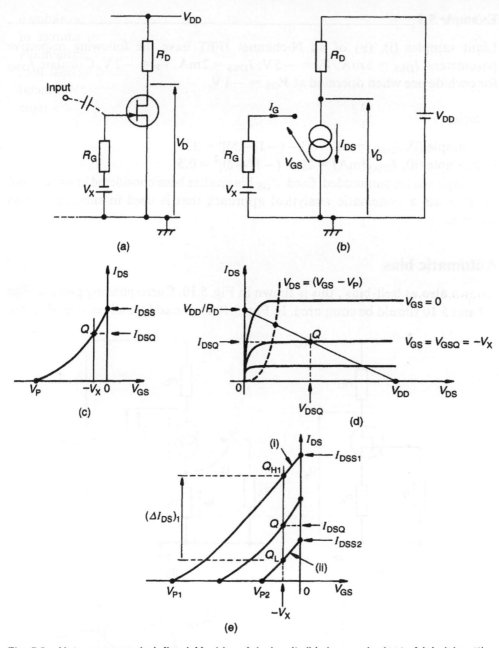

Fig. 5.9   Not recommended, fixed-V_{GS} bias: (a) circuit; (b) d.c. equivalent of (a); (c) setting the operating point Q; (d) Q point on V_{DS} curves; (e) effect of characteristic spreads on the location of Q

Fixed-V_{GS} biasing is unacceptable in practice because of the wide spread in device characteristics. This is shown, qualitatively, in (e), where Q_{H1} and Q_L are the high and low operating points that result from the limit chracteristics, (i) and (ii), respectively. The condition $(\Delta I_{DS})_1 > I_{DSQ}$ is possible.

Example 5.1

Limit samples (i), (ii) of an N-channel JFET have the following respective parameters: $I_{DSS} = 5\,mA$, $V_P = -5\,V$; $I_{DSS} = 2\,mA$, $V_P = -2\,V$. Calculate I_{DSQ} for each device when operated at $V_{GS} = -1\,V$.

Solution

For sample (i), $I_{DSQ}\,(mA) = 5[1-(-1/-5)]^2 = 3.2$
For sample (ii), $I_{DSQ}\,(mA) = 2[1-(-1/-2)]^2 = 0.5$
Although not recommended, fixed-V_{GS} biasing has been considered in some detail to illustrate a systematic analytical approach that is used in subsequent bias schemes.

Automatic bias

Known also as 'self-bias', this is shown in Fig. 5.10. Corresponding parts of Figs 5.9 and 5.10 should be compared. In Fig. 5.10(a), the additional resistor, R_S is the

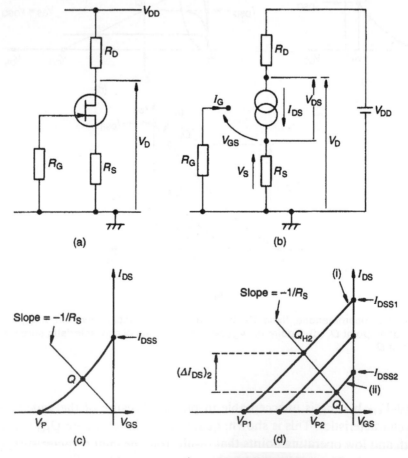

Fig. 5.10   'Automatic' or 'self' bias. $(\Delta I_{DS})_2 < (\Delta I_{DS})_1$ of Fig. 5.9(c)

crucial bias component. The FET produces its own bias by the flow of I_{DS} in this, hence the name. For the equivalent circuit in Fig. 5.10(b), ignoring I_G,

$$V_{GS} = V_G - V_S = -V_S = -I_{DS}R_S \qquad (5.12)$$

This provides a physical interpretation of circuit action.

If $R_S = 0$ then $I_{DS} = I_{DSS}$ and if $R_S = \infty$ then $I_{DS} = 0$, i.e. the source is open circuit. For $\infty > R_S > 0$, I_{DS} must self-limit at a value dependent on the precise value of R_S. Equation (5.12) can be re-written,

$$I_{DS} = -V_{GS}/R_S \qquad (5.13)$$

Equation (5.13) describes a gate-source circuit load line that passes through the origin and has a slope $-1/R_S$. This is shown, drawn on the typical mutual characteristic of Fig. 5.10(c), to give the quiescent point Q. The effect of characteristic spreads is shown in Fig. 5.10(d). For purposes of comparison with Fig. 5.9(e), R_S is assumed to be that value which gives the same Q_L with the low limit sample. Since the load line in Fig. 5.10(d) is not vertical, as it is in Fig. 5.9(e), it follows that, $(\Delta I_{DS})_2 < (\Delta I_{DS})_1$, as shown in Example 5.2 below.

Figure 5.10 does not include a sketch corresponding to Fig. 5.9(d) because no new principle is involved. Note, however, that the 'drain' d.c. load line now corresponds to $(R_D + R_S)$, because $V_{DS} = V_{DD} - I_{DS}(R_D + R_S)$. Hence, the load line passes through the axis points V_{DD}, $V_{DD}/(R_D + R_S)$.

Example 5.2

An N-channel JFET, having the limit characteristics of Example 5.1, is used in the circuit of Fig. 5.10(a). When the low current sample is used $I_{DSQ} = 0.5\,\text{mA}$ (i.e. that of the low current sample in Example 5.1). Calculate I_{DSQ} for the high current sample.

Solution

For the low current sample, $I_{DS}(\text{mA}) = 0.5 = 2[1 + (V_{GS}/2)]^2$

$$\therefore \ \sqrt{(1/4)} = 1 + (V_{GS}/2)$$

or,

$$V_{GS} = -1\,\text{V (as previously given)}$$

$$\therefore \ V_S = -V_{GS} = +1\,\text{V} = I_{DS}R_S$$

But, $I_{DS} = 0.5\,\text{mA}$,

$$\therefore \ R_S = 2\,\text{k}\Omega$$

In general, $V_{GS}(\text{V}) = -I_{DS}(\text{mA}) \times 2\,(\text{k}\Omega)$. Substituting this in the expression for the high current sample gives an equation for I_{DS} (expressed in mA).

$$I_{DS} = 5[1 - (-2I_{DS}/-5)]^2$$

or,

$$I_{DS} = 5[1 - (2I_{DS}/5)]^2$$

Expanding this expression gives the quadratic equation,

$$I_{DS} = 5[1 - (4I_{DS}/5) + (4I_{DS}^2/25)]$$

or,

$$I_{DS} = 5 - 4I_{DS} + (4I_{DS}^2/5)]$$

Hence,

$$(4I_{DS}^2/5) - 5I_{DS} + 5 = 0$$

Solving this gives $I_{DSQ} = 1.25\,\text{mA}$ and $I_{DSQ} = 5\,\text{mA}$. The solution $I_{DSQ} = 5\,\text{mA}$ cannot apply, because $I_{DS} = 5\,\text{mA}$ at $V_{GS} = 0\,\text{V}$, so the required value must be $I_{DSQ} = 1.25\,\text{mA}$. In this example $(\Delta I_{DS})_2 = 0.75\,\text{mA}$: for the previous example, $(\Delta I_{DS})_1 = 2.7\,\text{mA}$ so the advantage of automatic bias is evident.

The reason for two values of I_{DSQ} in Example 5.2 and similar problems is explained by the two points Q and Q' in Fig. 5.11. The equation for the transfer characteristic describes a complete parabola with its apex located on the V_{GS} axis at V_P and thus includes the dotted section that has no physical reality: only that part of the parabola corresponding to $(0 \geq)\ V_{GS} > V_P$, and having a positive slope, relates to the physical electronics of device operation. The relevant I_{DSQ} is given by the insection point Q, i.e. the *smaller* of the two calculated values.

Automatic bias is not possible with BJTs because $I_C = 0$ when $V_{BE} = 0$. Manufacturers supply FETs strapped to operate as 'constant-current diodes' for

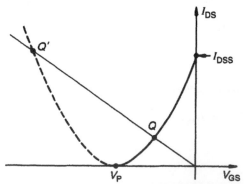

Fig. 5.11 Phantom parabola section (dotted) gives illusory operating point, Q', in FET calculations

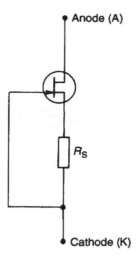

Fig. 5.12   Constant-current 'diode'

use in biasing zener diodes, etc. The circuit arrangement is shown in Fig. 5.12. R_S can be trimmed, during manufacture, to meet required I_{DS} specifications: the choice $R_S = 0$ is acceptable when close limits on I_{DS} are not required. In all cases it is necessary to operate with $V_{AK} > V_P$.

Gate-potentiometer bias

This is analogous to the base-potentiometer bias technique used with BJTs and is shown in Fig. 5.13. It differs from automatic bias in that it uses a positive gate voltage, V_{GG}, derived from V_{DD} by a potential divider chain comprising R_{G1}, R_{G2}. Labelled sections of Fig. 5.13, when compared with their counterparts in Fig. 5.10, help to explain circuit operation.

Referring to Fig. 5.13(b), the section inside the dotted contour is a Thévenin equivalent circuit of the gate potentiometer network. Hence,

$$R_{GG} = R_{G1} \| R_{G2} = R_{G1} R_{G2} / (R_{G1} + R_{G2}) \tag{5.14}$$

and,

$$V_{GG} = V_{DD} R_{G1} / (R_{G1} + R_{G2}) \tag{5.15}$$

Ignoring I_G,

$$V_S = I_{DS} R_S = V_{GG} - V_{GS} \tag{5.16}$$

$$\therefore I_{DS} = -(V_{GS}/R_S) + (V_{GG}/R_S) \tag{5.17}$$

Equation (5.17) describes the gate-source load line which is plotted on the typical transfer characteristic in Fig. 5.13(c).

If V_{GG} and R_S are chosen so that Q_L (in Fig. 5.13(d)) is the same as in the two previous bias schemes then $(\Delta I_{DS})_3 < (\Delta I_{DS})_2$, as shown in the next example. For

The load line also cuts the low limit characteristic, (ii), at Q_2 corresponding to $V_{GS} = V_{GS2}$ (which can, likewise, be calculated) and cuts the line $I_{DS} = I_2$ at P_2, for which $V_{G2} = V_2$. From the geometry of the diagram,

$$V_1 = V_{GG} - I_1 R_S \tag{5.18a}$$

$$V_2 = V_{GG} - I_2 R_S \tag{5.18b}$$

To be acceptable the load line must be positioned, as shown, so that,

$$V_{GS1} \geq V_1 \tag{5.19a}$$

and

$$V_{GS2} \leq V_2 \tag{5.19b}$$

Substituting for V_1, V_2 from the Equations (5.18) we obtain,

$$V_{GG} \leq V_{GS1} + I_1 R_S \tag{5.20a}$$

$$V_{GG} \geq V_{GS2} + I_2 R_S \tag{5.20b}$$

Consider now Fig. 5.15. On a plot of V_{GG} vs R_S, the use of the equality sign in Equation (5.20a) gives rise to the straight line A. Consequently, Equation (5.20a) is satisfied for all pairs of values of V_{GG} and R_S, that lie on or below line A.

Similarly, Equation (5.20b) is satisfied for all pairs of values of V_{GG} and R_S, that lie on or above line B. Thus, the only acceptable pairs of values of V_{GG} and R_S, are those which lie in the shaded area between A and B.

A suitable operating point, Q, is one which lies inside a rectangle that defines tolerances on V_{GG} and R_S, and is itself, completely in the shaded region. Once V_{GG} and R_S are chosen it remains to find R_{G1} and R_{G2}.

From Equations (5.14) and (5.15) it follows that,

$$R_{G1} = V_{DD} R_{GG} / (V_{DD} - V_{GG}) \tag{5.21}$$

$$R_{G2} = V_{DD} R_{GG} / V_{GG} \tag{5.22}$$

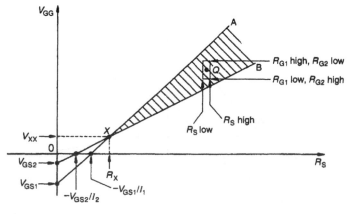

Fig. 5.15   Shaded region shows acceptable operating area

Table 5.2   Step-by-step design procedure for toleranced gate-potentiometer bias circuit

Step	Procedure
1	Calculate $V_{GS1} = V_{P1}(1 - \sqrt{I_1/I_{DSS1}})$
2	Calculate $(-V_{GS1}/I_1)$
3	Plot line A (*see* Fig. 5.14) from intercepts in 1, 2
4	Calculate $V_{GS2} = V_{P2}(1 - \sqrt{I_2/I_{DSS2}})$
5	Calculate $(-V_{GS2}/I_2)$
6	Plot line B (*see* Fig. 5.14) from intercepts in 4, 5
7	Select R_S (preferred value), V_{GG}, in operating area
8	Calculate $R_{G1} = V_{DD}R_{GG}/(V_{DD} - V_{GG})$
9	Calculate $R_{G2} = V_{DD}R_{GG}/V_{GG}$
10	Select preferred values of R_{G1}, R_{G2}
11	Check that V_{GG}, R_S are still in operating area when tolerances are included

Note that if the shaded region in Fig. 5.15 encloses a section of the R_S axis then R_{G2} can be dispensed with and the problem can be solved using automatic bias. V_{XX} and R_X correspond to a load line (Fig. 5.14) that passes through operating points at the intersection of line $I_{DS} = I_1$ with curve (i) and line $I_{DS} = I_2$ with curve (ii). This load line is not acceptable because of resistor tolerances. A design procedure, based on the discussion given, is shown on Table 5.2. Steps 7–11, inclusive, may need repetition to ensure that the operating point remains in the shaded zone of Fig. 5.15 when resistor tolerances (and a possible restriction on the maximum allowable V_{GG}) are taken into account. If R_{GG} is large, e.g. $R_{GG} \geq 10\,\mathrm{M\Omega}$, to meet an input resistance requirement, application of Equations (5.21) and (5.22) may lead to excessively large values of R_{G1} and R_{G2}. A useful procedure then is to use the modified gate circuit of Fig. 5.16. R_1 and R_2, are chosen to give the required V_{GG} and may be in the kilohm range. R_3 is then selected to give the effective value, R'_{GG}, of bias resistor, as seen from the gate terminal: $R'_{GG} = (R_{GG} + R_3) = [(R_1||R_2) + R_3]$.

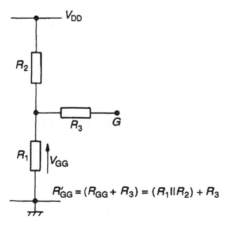

Fig. 5.16   Alternative bias scheme when R'_{GG} is required to have a high resistance

N.B. In each of the following problems assume that the FET used may have the limit characteristics used in Example 5.1 ($I_{DSS} = 5\,\text{mA}$, $V_P = -5\,\text{V}$; $I_{DSS} = 2\,\text{mA}$, $V_P = -2\,\text{V}$).

9   In the voltage-reference circuit of Fig. 5.20, Q_F supplies a constant current to Q_B, which is strapped to operate as a zener diode. The base–emitter junction of Q_B, in the breakdown region, can be modelled by an 8.2 V battery with internal resistance 10 Ω. When Q_F is the high limit sample, $V_Z = V_{Z1}$: for the low current sample $V_Z = V_{Z2}$. Calculate $(V_{Z1} - V_{Z2})$ and determine the minimum value of V_{DD}, rounded-up to the nearest 1 V, for Q_F to operate as a current source.

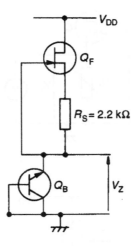

Figure 5.20

10   Find I_{DSQ} (max) and I_{DSQ} (min) for the circuit of Fig. 5.21.

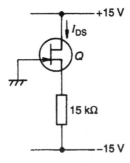

Figure 5.21

11   In the source-current bias circuit of Fig. 5.22 the current-mirror transistors Q_{B1} and Q_{B2} may be assumed to have identical characteristics, namely,

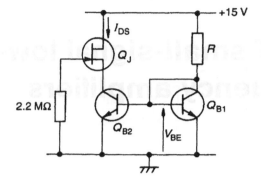

Figure 5.22

$\beta_0 = 100$, $V_A = \infty$, and to operate at $V_{BE} = 0.7\,\text{V}$. Show that $R > 16.6\,\text{k}\Omega$ if Q_{B2} is to operate in the forward-active mode.

12 Calculate $I_{DSQ}(\text{max})$ and $I_{DSQ}(\text{min})$ for the circuit of Fig. 5.23.

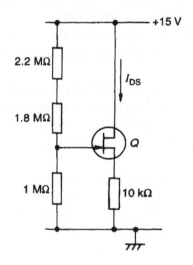

Figure 5.23

13 Design a fully-toleranced gate-potentiometer bias scheme to give $0.75\,\text{mA} > I_{DS} > 0.5\,\text{mA}$ when $V_{DD} = 15\,\text{V}$, $V_{GG}(\text{max}) < 5\,\text{V}$, and $R'_{GG} = 10\,\text{M}\Omega \pm 6\%$. (Hint: use the design procedure of Table 5.2 and the gate circuit of Fig. 5.16.)

corresponds to $I_{DS} = 0$ and $V_{DS} = V_{DD}$. The upper limit is set by the point Q_X, at the intersection of (i) with the straight line (ii), which is a plot of the condition $V_{DS} = (V_{GS} - V_P)$.

The transfer characteristic that lies above this may be considered linear for *small* changes about a quiescent point Q. The meaning of 'small' becomes evident from the discussion in the next section.

6.2 LF EQUIVALENT CIRCUIT OF THE JFET

The procedure for deriving the incremental model of a JFET from the modified square-law model, by a consideration of the slopes of the characteristics at the operating point, is similar to that used to derive the l.f. incremental model of a BJT.

Derivation of equivalent circuit

Figure 6.2(a) shows a d.c. representation of an N-channel JFET. Figure 6.2(b) shows expanded views of sections of its transfer and output characteristics in the vicinity of a chosen Q-point, characterized by the quantities V_{GSQ}, V_{DSQ}, I_{DSQ}:v_{gs}, v_{ds}, i_{ds} represent changes from these d.c. conditions. As a matter of convention, they are all considered to be positive increments in the derivation of the equivalent circuit. Note, however, that in circuit operation the condition $v_{gs} > 0$ normally gives $v_{ds} < 0$. Assuming v_{gs}, v_{ds} and i_{ds} are small we can approximate the sections of characteristic by tangents at the Q-point. The input appears as an incremental open circuit because $V_{GSQ} < 0$.

At the output the change i_{ds} can be considered to be the sum of two components, i_1 and i_2. The component i_1 results from the change v_{gs} with V_{DS} held constant at $V_{DS} = V_{DSQ}$; i_2 results from the change v_{ds}, V_{GS} being kept constant at $V_{GS} = V_{GSQ} + v_{gs}$. Thus,

$$i_1 = g_{fs}v_{gs} \qquad (6.1)$$

where, g_{fs} (or g_m) is the mutual conductance, or transconductance: g_{fs} = slope of I_{DS} against V_{GS} curve, for $V_{DS} = V_{DSQ}$, at $V_{GS} = V_{GSQ}$.

The subscripts f and s refer, respectively, to forward (as in 'forward transmission' of a signal through the JFET) and source (the terminal being reckoned as common). Similarly,

$$i_2 = v_{ds}/r_{ds} \qquad (6.2)$$

where, r_{ds} ($= 1/g_{ds}$) is the incremental output resistance: r_{ds} = inverse of slope of I_{DS} against V_{DS} curve, for $V_{GS} = (V_{GSQ} + v_{gs}) \simeq V_{GSQ}$, at $V_{DS} = V_{DSQ}$. Subscripts d and s refer to the drain and source terminals.

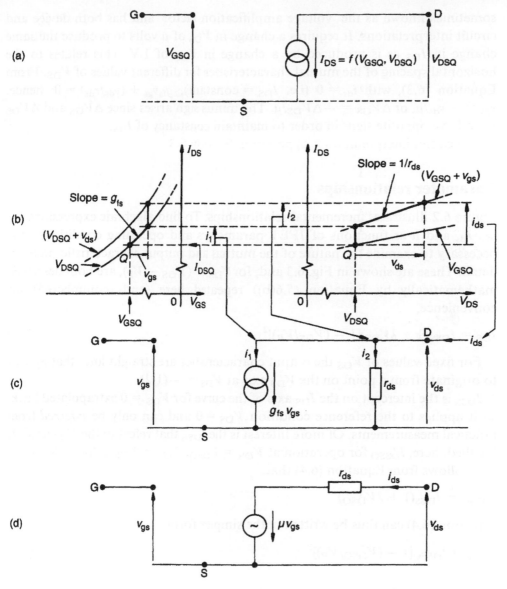

Fig. 6.2. Derivation of l.f. incremental model of a JFET. (a) D.C. representation. (b) Operation in vicinity of Q-point. (c) VCCS derived from (b). (d) VCVS derived from (c)

The assumption $v_{gs} \ll V_{GSQ}$ implies $i_1 \ll I_{DSQ}$. Since $i_{ds} = (i_1 + i_2)$ it follows from Equations (6.1) and (6.2) that,

$$i_{ds} = g_{fs}v_{gs} + (v_{ds}/r_{ds}) \qquad (6.3)$$

The incremental equivalent circuit described by Equation (6.3) is shown in Fig. 6.2(c): it is a VCCS with a finite output resistance.

An alternative VCVS, with finite output resistance, that is equivalent is shown in Fig. 6.2(d). It is derived from Fig. 6.2(c) by use of the Thévenin–Norton transformation. Circuit equivalence requires that $\mu = g_{fs}r_{ds}$. The parameter μ is sometimes known as the 'voltage amplification factor' and has both device and circuit interpretations. It requires a change in V_{DS} of μ volts to produce the same change in I_{DS} as is produced by a change in V_{GS} of 1 V. This relates to the horizontal spacing of the mutual characteristics for different values of V_{DS}. From Equation (6.3), with $i_{ds} = 0$ (i.e. $I_{DS} =$ constant), $g_{fs}v_{gs} + (v_{ds}/r_{ds}) = 0$: hence, $v_{gs} = -v_{ds}/\mu$, or $\Delta V_{GS} = -\Delta V_{DS}/\mu$. The minus sign arises since ΔV_{GS} and ΔV_{DS} must have opposite signs in order to maintain constancy of I_{DS}.

The circuit interpretation is apparent in Section 6.3.

Parameter relationships

Figure 6.2 illustrates incremental relationships. To find algebraic expressions for g_{fs}, r_{ds} and μ as functions of JFET parameters and operating conditions, it is necessary to examine the nature of the mutual and output characteristics in more detail. These are shown in Fig. 6.3 and, for $V_{DS} > (V_{GS} - V_P)$, are both described mathematically by Equation (5.6(b)) repeated here (and re-numbered) for convenience.

$$I_{DS} = I_{DSS}(1 + \lambda V_{DS})[1 - (V_{GS}/V_P)]^2 \tag{6.4}$$

For fixed values of V_{GS} the output characteristics are straight lines that appear to originate from a point on the V_{DS} axis at $V_{DS} = -(1/\lambda)$.

I_{DSS} is the intercept on the I_{DS} axis of the curve for $V_{GS} = 0$ extrapolated back, so it applies to the reference condition $V_{DS} = 0$ and can only be *inferred* from practical measurements. Of more interest is the I_{DSS} that refers to the V_{DS} used. It is called, here, I_{DSSQ} for operation at $V_{DS} = V_{DSQ}$, $V_{GS} = V_{GSQ}$, $I_{DS} = I_{DSQ}$.

It follows from Equation (6.4) that,

$$I_{DSSQ} = I_{DSS}(1 + \lambda V_{DSQ}) \tag{6.5}$$

Equation (6.4) can thus be written in the simpler form,

$$I_{DSQ} = I_{DSSQ}[1 - (V_{GSQ}/V_P)]^2 \tag{6.6}$$

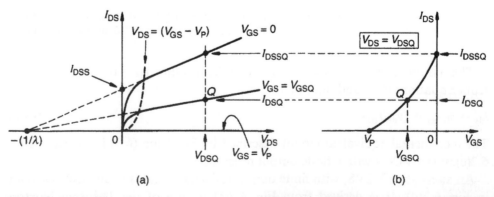

Fig. 6.3 I_{DSSQ} defined on the output characteristics, (a), and the transfer characteristic, (b), for $V_{DS} = V_{DSQ}$

Equation (6.5) describes the section of parabola (Fig. 6.3(b)) that comprises the mutual characteristic for $V_{DS} = V_{DSQ}$.

Consider first r_{ds}. From the geometry of Fig. 6.3(a), the slope of the output characteristic for $V_{GS} = V_{GSQ}$, at Q is,

$$(1/r_{ds}) = I_{DSQ}/[V_{DSQ} + (1/\lambda)] \tag{6.7}$$

For $(1/\lambda) \gg V_{DSQ}$ this means,

$$r_{ds} \simeq 1/(\lambda I_{DSQ}) \tag{6.8}$$

As with the BJT, the incremental output resistance is inversely related to operating current.

If $(1/\lambda)$ is expressed in volts and I_{DSQ} in mA, then r_{ds} is given in kΩ. As the output characteristics are assumed straight, r_{ds} is independent of v_{ds}, i.e. it appears that there is no 'small-signal' restriction. However, this assumption is only approximately valid because in practice the characteristics are somewhat curved. As a rule-of-thumb, r_{ds} can be assumed constant for $v_{ds} < 5$ V.

Consider next g_{fs}. This can be found from Equation (6.4) or (6.6) by differentiation (*see* Problem 1), but that method fails to resolve the meaning of 'small-signal' as far as the input is concerned so we proceed as follows.

For an increment v_{gs} in V_{GSQ} and i_1 in I_{DSQ}, with V_{DS} constant at V_{DSQ}, Equation (6.6) becomes,

$$I_{DSQ} + i_1 = I_{DSSQ}[1 - \{(V_{GSQ} + v_{gs})/V_P\}]^2 \tag{6.9}$$

Subtracting Equation (6.6) from (6.9),

$$i_1 = I_{DSSQ}[A^2 - B^2] \tag{6.10a}$$

or,

$$i_1 = I_{DSSQ}[A - B][A + B] \tag{6.10b}$$

In Equations (6.10), $A = [1 - \{V_{GSQ} + v_{gs}\}/V_P\}]$ and $B = [1 - (V_{GSQ}/V_P)]$. Hence,

$$i_1 = I_{DSSQ}(-v_{gs}/V_P)[2 - 2(V_{GSQ}/V_P) - (v_{gs}/V_P)] \tag{6.11}$$

The third term inside the square brackets can be ignored if,

$$|(v_{gs}/V_P)/[2 - 2(V_{GSQ}/V_P)]| < 0.1 \tag{6.12}$$

or,

$$|v_{gs}| \leqslant |(V_P - V_{GSQ})|/5 \tag{6.13}$$

Then, $i_1 \propto v_{gs}$, or $i = g_{fs} v_{gs}$ where,

$$g_{fs} = -2(I_{DSSQ}/V_P)[1 - (V_{GSQ}/V_P)] \tag{6.14}$$

Relationship (6.13) provides a practical definition for 'small' v_{gs} for given d.c. conditions.

Suppose $V_P = -3$ V, $V_{GSQ} = -1.5$ V. Then 'small' means $|v_{gs}| < 0.3$ V. This is much greater than the small input-signal requirement for a BJT (i.e. $v_{be} \leqslant 5$ mV). Eliminating $[1 - (V_{GSQ}/V_P)]$ between Equations (6.6) and (6.14),

$$g_{fs} = -(2/V_P)\sqrt{I_{DSQ}I_{DSSQ}} \tag{6.15}$$

For V_P expressed in volts and I_{DSQ}, I_{DSSQ} in mA, g_{fs} is expressed in mS, the usual unit for low power JFETs.

Despite the appearance of the minus sign in Equation (6.15), $g_{fs} > 0$. Thus, for an N-channel JFET, $V_P < 0$ and the plus sign must be selected in taking the square root because $I_{DSQ} > 0$ and $I_{DSSQ} > 0$. Hence $g_{fs} > 0$, as it must be because I_{DS} increases with V_{GS}.

Equation (6.15) is a convenient form for g_{fs} because, in the design of a bias circuit, attention is directed to a required I_{DSQ}. However, an alternative form for g_{fs}, that is illuminating, is obtained as follows. At $V_{GS} = 0$, $g_{fs} = g_{fso}$ and is given by putting $I_{DSQ} = I_{DSSQ}$ in Equation (6.15).

$$\therefore \quad g_{fso} = -2I_{DSSQ}/V_P \tag{6.16}$$

From Equations (6.14) and (6.16) it follows that,

$$g_{fs} = g_{fso}[1 - (V_{GSQ}/V_P)] \tag{6.17}$$

This relationship is shown plotted in Fig. 6.4. The control of g_{fs}, via the d.c. level of V_{gs}, is made use of in automatic gain control (a.g.c.) circuits in amplifier systems.

An expression for μ is obtained from Equations (6.8) and (6.15). Thus,

$$\mu = g_{fs}r_{ds} = (2/\lambda V_P)\sqrt{I_{DSSQ}/I_{DSQ}} \tag{6.18}$$

On JFET data sheets, g_{fs} is often specified as the magnitude of the transfer admittance at some low-frequency of measurement, e.g. 1 kHz, and specified d.c.

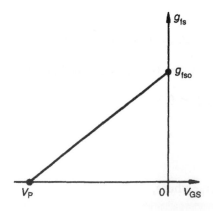

Fig. 6.4   Variation of g_{fs} with V_{GS}

conditions and written $|y_{fs}|$. Similarly, $(1/r_{ds})$ is usually given as $|y_{os}|$. From this $(1/\lambda)$ must be inferred as it is not usually given. A rule-of-thumb figure for $(1/\lambda)$ when $|y_{os}|$ is not known, or specified, is 100 V.

The high frequency JFET model

Figure 6.5 shows a high frequency (h.f.) 'π-model' of the JFET. It is Fig. 6.2(c) with the addition of C_{gs}, C_{gd}, C_{ds}, which take into account internal and external capacitive effects between the electrodes of the device.

Data sheets quote either C_{gs}, C_{gd}, or the capacitance parameters, C_{gss} and C_{dss}, from which they can be deduced. Thus,

$$C_{gss} \text{ (or } C_{iss}) = (C_{gs} + C_{gd}) \tag{6.19}$$

and,

$$C_{dss} \text{ (or } C_{oss}) = (C_{gd} + C_{ds}) \simeq C_{gd} \tag{6.20}$$

The approximation in Equation (6.20) holds because $C_{ds} \ll C_{gd}$. (C_{gd} is also known as C_{rss}.) Hence, from Equations (6.19) and (6.20),

$$C_{gs} = (C_{iss} - C_{gd}) = (C_{iss} - C_{rss}) \tag{6.21}$$

and,

$$C_{gd} = C_{rss} \simeq C_{oss} \tag{6.22}$$

Although C_{gd} may be smaller than C_{gs} it has a dominating effect in h.f. amplification, as is shown in Chapter 7.

For the present, 'low-frequency' means that frequency range over which the existence of the capacitances can be ignored because the currents in them are negligible compared with those in the other elements in the model.

Before leaving the topic of equivalent circuits, note that to avoid the tedious repetition of subscripts, it is convenient to write v_{gs} as v (i.e. no subscripts) and v_{ds}, where appropriate, as v_o. For calculations involving sinusoidal signals, v is written in RMS form as V, and v_o as V_o.

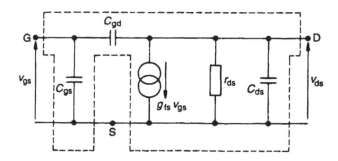

Fig. 6.5 High-frequency π-model of the JFET

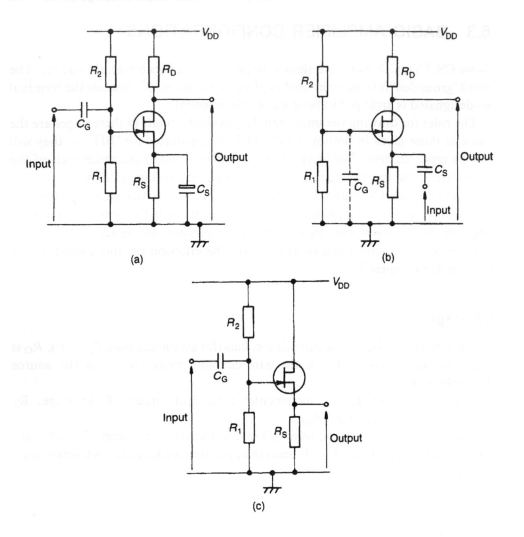

(a)

(b)

(c)

Fig. 6.6   Basic amplifier stages: (a) CS; (b) CG; (c) CD

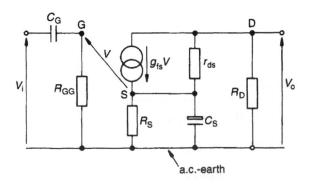

a.c.-earth

Fig. 6.7   Small-signal equivalent circuit of Fig. 6.6(a) for sinusoidal input. $R_{GG} = R_1 \| R_2$

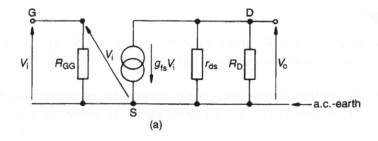

(a)

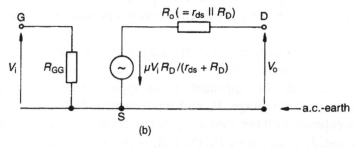

(b)

Fig. 6.8 (a) Simplified form of Fig. 6.7 (C_G, C_S regarded as a.c. short-circuits). (b) VCVS form of (a)

The operation of the stage is illustrated graphically in Fig. 6.9. The d.c. load line is described by the relationship $I_{DS} = (V_{DD} - V_{DS})/(R_D + R_S)$. However, when C_S is present, the output excursion is defined by the a.c. load line: this is given by $I_{ds} = -V_{ds}/R_D$ because R_S is decoupled. The calculation of current gain, A_i, and power gain, A_p, is left as an exercise for the reader (*see* Problem 8).

The CS stage finds general application as an l.f. audio frequency pre-amplifier/ amplifier.

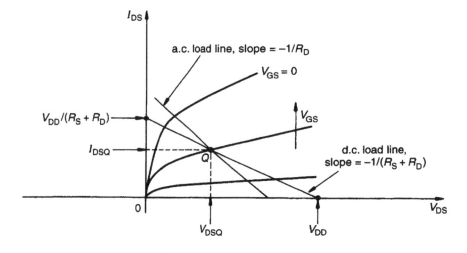

Fig. 6.9 Showing the a.c. load line for Fig. 6.6(a)

The discussion, above, illustrates the widely-exploited resistance-transformation properties of the stage.

Looking-in at the source, resistances in series with the drain terminal appear to be reduced by a factor $(\mu + 1)$, typically two orders of magnitude: looking-in at the drain, resistances in series with the source terminal appear to be increased by the same factor.

Example 6.3 —————————————————————————————

Calculate the magnitude of the current gain, A_i, for the circuit of Fig. 6.6(b) given the following data: $R_S = 1\ k\Omega$; $R_D = 4.7\ k\Omega$; $r_{ds} = 50\ k\Omega$; $g_{fs} = 2\ mS$. C_S may be considered to have zero reactance.

Solution

Referring to Fig. 6.10, the input current I_i is split between R_S and the incremental resistance seen looking-in at the source terminal, i.e. $(R_D + r_{ds})/(\mu + 1)$. Hence, by the current-splitting rule,

$$|A_i| = R_S/[R_S + \{(R_D + r_{ds})/(\mu + 1)\}]$$

From the data given, $\mu = 2\ mS \times 50\ k\Omega = 100$. Thus, $|A_i| = 1/[1 + (54.7/100)]$ or, $|A_i| \simeq 0.65$.

CD stage (the 'source-follower' circuit)

Figure 6.6(c) shows an elementary form of this but there are many variations, one of which is discussed briefly later.

In the CD stage, there is a source resistor but no source-decoupling capacitor. Consequently, the potential of the source, from which an output is taken, is able to follow that of an input signal applied between the gate terminal and a.c.-earth: hence, the name 'source-follower'.

A small-signal equivalent circuit of Fig. 6.6(c) is shown in Fig. 6.12 and the analysis of it follows, closely, that used for the CG stage. By inspection,

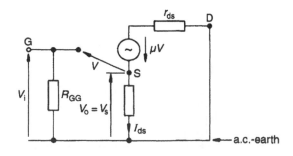

Fig. 6.12   Simplified equivalent of Fig. 6.6(c)

$R_i = R_{GG}$. I_{ds} is given by the output loop equation,

$$I_{ds}(R_D + r_{ds} + R_S) = \mu V \tag{6.43}$$

But,

$$V = (V_i - V_s) = V_i - I_{ds}R_S \tag{6.44}$$

Hence,

$$I_{ds}(R_D + r_{ds} + R_S) = \mu(V_i - I_{ds}R_S) \tag{6.45}$$

or,

$$I_{ds} = \mu V_i/[r_{ds} + (\mu + 1)R_S] \tag{6.46}$$

and,

$$A_v = (V_o/V_i) = \mu R_S/[r_{ds} + (\mu + 1)R_S] \tag{6.47}$$

If $(\mu + 1)R_S \gg r_{ds}$ and $\mu \gg 1$, Equation (6.47) gives,

$$A_v \simeq 1 \tag{6.48}$$

Looking-in at the source we see the incremental input resistance of the CG stage but with $R_D = 0$. Hence, making use of the Equation (6.34),

$$(1/R_o) = [(1/R_S) + (1/r_{ds}) + g_{fs}] \tag{6.49}$$

Since, in practice, $r_{ds}, R_S \gg (1/g_{fs})$,

$$R_o \simeq (1/g_{fs}) \tag{6.50}$$

The source-follower is often called a 'unity-gain buffer stage' and it finds application wherever a high or very high input resistance is required. A variation, which is at the core of the design of an oscilloscope probe that has a high input inpedance over a wide frequency range, is shown in Fig. 6.13. It is an elegant

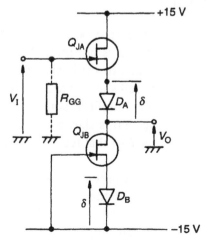

Fig. 6.13 A variation of the basic 'source-follower'

(d) **Square-law transfer characteristic**

The square-law relationship between I_{DS} vs V_{GS} facilitates the design of a range of modulation and mixing circuits in telecommunications.

(e) **Resistance to radiation effects**

Operation of the BJT relies on the transport of minority carriers and that is significantly affected by incident electromagnetic radiation. (Hence, the existence of photo-diodes and photo-transistors, etc.)

The FET depends on the transport of majority carriers, which is not significantly affected by radiation. This makes the use of FETs attractive in environments where there is likely to be nuclear radiation.

(f) **Construction** (applicable to MOSFETs)

Single-polarity MOSFETs are popular in IC structures because they require no PN-junctions to isolate them. This is because all the essential electrical activity is confined to the region under a gate. Thus, small-area large-scale stores (ROMs, RAMs) can be constructed from basic logic schemes. Note, however, that isolation *is* required in CMOS IC structures, where two polarities of device are employed.

BJTs are superior to FETs with respect to predictability of transfer characteristic. As mentioned in Chapter 3, the exponential relationship between I_C and V_{BE} is accurately valid over many decades of collector current and this is the basis on which precision bipolar analogue ICs are designed.

The nature of the relationship between I_C and V_{BE} means that g_m is accurately defined, if I_C and T are known, and comparatively large (~ 40 mS per mA of I_C). Consequently, CE amplifier gain ($\simeq -g_m R_C$) is reasonably well-defined. In the case of the FET, the transfer characteristic is subject to wide manufacturing tolerances and the assumption that it is square-law in nature is only a useful engineering approximation. Consequently, g_{fs} is poorly defined: furthermore, it is only modest in magnitude, e.g. 5 mS when I_{DS} is several mA.

With respect to operating voltages and frequency range it is difficult to make non-contentious statements because of continual developments in device technology. BJTs can operate with rail voltages of 1.5 V (in hearing-aids). FETs generally need more than this. The voltage swing required to switch-on BJTs in certain types of logic circuit (e.g. emitter-coupled logic, ECL) can be as low as 0.25 V and this results in operation up to, and beyond, a clock frequency of 1 GHz (i.e. 1000 MHz). Gallium arsenide FET logic circuits can match this operating frequency but the devices are more difficult to make than BJTs.

In conclusion, efficient circuit design combines the use of BJTs and FETs, when this is appropriate, so that the unique properties of each device type are exploited to the full.

6.5 SELF ASSESSMENT TEST

1 By what factor does the g_{fs} of a JFET change when the quiescent drain current doubles?

2 Define the parameter μ of a JFET.

3 A JFET in a CS stage has $V_P = -4\,V$ and operates at $V_{GS} = V_{GSQ} = -2\,V$. Specify an upper limit to the amplitude of a signal, applied between gate and source, if it is to be considered small enough for the l.f. incremental circuit to be valid.

4 Why is the power gain of the CS (and CD) configuration very high?

5 Explain, in words, why the incremental resistance looking into the drain of an N-channel JFET, in the CS configuration, increases when the source bypass capacitor is removed.

6 Explain why a CS stage and a CG stage, operating at the same I_{DS} and with the same R_D, have effectively the same magnitude of voltage gain when driven from a generator with zero output resistance.

7 Name an application area for which each of the basic JFET configurations (CS, CG, CD) is well suited.

8 State three application areas for which the use of a FET is preferable to that of a BJT.

6.6 PROBLEMS

1 Obtain Equations (6.7) and (6.15) by differentiating Equation (6.4).

2 Show that the tangent to an N-channel JFET transfer characteristic at $V_{GS} = 0$ intersects the V_{GS} axis at $V_{GS} = V_P/2$. (This fact is the basis of a measurement technique for V_P.)

3 Obtain Equation (6.17) from Equations (6.14) and (6.16).

3 Show that $(g_{fs}/g_{fso}) = \sqrt{I_{DSQ}/I_{DSSQ}}$.

4 Sketch, roughly, to a common scale of I_{DS}:
   (a) g_{fs} vs I_{DS};   (b) r_{ds} vs I_{DS};   (c) μ vs I_{DS}.

5 An N-channel JFET has the specification given in Example 6.1. Calculate g_{fs}, r_{ds} and μ when the device is operated at $V_{DSQ} = 20$ V, $V_{GSQ} = -2$ V.

6 At $V_{GS} = 0$, $V_{DS} = 15$ V an N-channel JFET gave $I_{DS} = 4$ mA and measurements at 1 kHz, with this bias condition, gave $|y_{os}| = 40\ \mu$S, $|y_{fs}| = 4$ mS. Calculate r_{ds}, g_{fs} at $V_{DSQ} = 20$ V, $I_{DSQ} = 2$ mA.

7 For the circuit of Fig. 6.6(a): $R_1 = 1$ MΩ; $R_2 = \infty$; $R_S = 1$ kΩ; $R_D = 10$ kΩ; $V_{DD} = 20$ V. The JFET has $V_P = -2$ V and its output characteristics are specified by the following data at $V_{GS} = 0$ V, $V_{DS} = 10$ V: $I_{DS} = 4$ mA; $g_{ds} = 40\ \mu$S. Calculate A_v, assuming the reactances of C_G, C_S are zero.

8 Show that A_i, A_p, of the CS stage are $A_i = \mu R_{GG}/(r_{ds} + R_d)$, $A_p = \mu^2 R_{GG}R_D/(r_{ds} + R_D)^2$.

9 Calculate (V_o/V_g) for the circuit of Fig. 6.15. Assume C_S has zero reactance and the JFET parameters are $g_{fs} = 3$ mS, $r_{ds} = 50$ kΩ.

10 Show that for Fig. 6.6(c), $A_p \simeq A_i \simeq (R_{GG}/R_S)$. (Make reasonable assumptions)

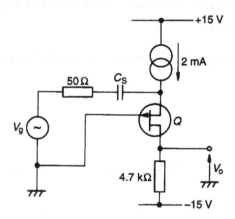

Figure 6.15

11 Derive Equation (6.54) from Equation (6.47). (Neglect r_e, the incremental resistance of each diode in Fig. 6.13.)

12 Derive Equation (6.56) by applying the condition $V_{DG} \geqslant -V_p$ to Q_{JA}, Q_{JB} in Fig. 6.13.

13 Show that for the circuit of Fig. 6.14,

$$(V_o/V_i) = -\mu R_D/[r_{ds} + R_D(\mu + 1)R_{S1}]$$

$$(V_o/V_s) = -R_D/R_{S1}$$

(Assume C_G and C_S have zero reactances.)

14 In Fig. 6.16, the moving-coil meter M and associated circuitry constitute a d.c. voltmeter with a high input resistance. Adjustment of R_V permits M to indicate a zero reading with $V_I = 0$, when the monolithic JFET pair Q_A, Q_B exhibit a V_{GS} match lying in the range ± 5 mV. Calculate suitable values of R_X and R_V.

Estimate the percentage error in the reading of M for $V_I \neq 0$, after the zero-setting has been made, assuming the resistance of M is 100 kΩ and Q_A, Q_B each have a g_{fs} of 2 mS. (Hint: assume the incremental output resistance at each source is $1/g_{fs}$.)

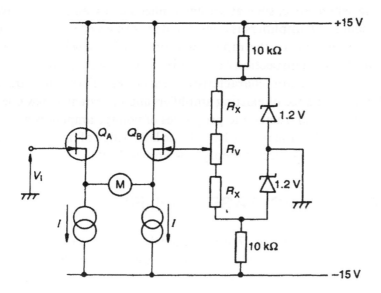

Figure 6.16

(f)), that are related to Fig. 7.1(d) together constitute the 'frequency response' of the gain A. Sometimes the expression is used for the single plot $|A|$ vs ω, or f.

Suppose that, instead of a sinusoidal signal, the generator produces the 'square' waveform shown in Fig. 7.2(a). This appears frequently in pulse electronics and is used in the testing of amplifiers and the adjustment of oscilloscope probes.

There is a theorem due to the French mathematician Fourier that is of fundamental importance in analogue systems. In essence, Fourier's Theorem states that any continuous repetitive waveform that repeats itself after a (periodic) time interval T can be expressed as the sum of a d.c. level and a series of sinusoidal signals known as 'frequency components'. Each component has a frequency that is an integral multiple of the 'fundamental' frequency ω ($= 2\pi f = 2\pi/T$), which is also known as the 'first harmonic'. The amplitudes of the components and the times at which their peaks occur, relative to the fundamental, can be determined by an established mathematical procedure, given in books on mathematics and on circuit theory (Hayt & Kemmerly, 1993).

For the particular waveform in Fig. 7.2(a) the d.c. level is zero and only the odd harmonics are present.

$$v_g(t) = (4V_m/\pi)[\sin \omega_1 t + (1/3) \sin 3\omega_1 t + (1/5) \sin 5\omega_1 t \ldots] \qquad (7.3)$$

Figure 7.2(b) shows the result of combining the first three terms in this expression.

Theoretically, an infinite number of frequency components is required to produce a faithful copy of the square wave but the amplitudes of the components decrease with frequency so, in practice, it is sufficient to approximate the waveform by a limited number.

When plotted to a base of frequency, the magnitude and phase of the frequency components comprise the 'line spectra' of the waveform. Figure 7.2(c) shows the magnitude spectrum such as might be observed on the display of a test instrument known as a 'spectrum analyzer': only the first three components are shown.

Consider now what happens when v_g in Fig. 7.1(a) is the waveform shown in Fig. 7.2(a). As the amplifier is linear the Principle of Superposition applies. This means that each frequency component of the input can be considered separately and that the resultant amplifier output is the sum of the contributions due to each of them.

Distortionless amplification is achieved with the amplifier in question because each of the components is magnified by the same numerical factor and experiences zero phase change in passing through the amplifier. Hence, the output is a magnified copy of the input square wave.

The general conditions for distortionless amplification, irrespective of input waveshape, are as follows. First, $|A|$ must be independent of signal amplitude, otherwise there is non-linear, or amplitude, distortion. Second, $|A|$ must be frequency-independent, otherwise frequency-distortion occurs. Third, to avoid phase-distortion, there are three choices for the amplifier phase shift, ϕ: $\phi = 0$ for all ω: $\phi = \pi$ for all ω: $\phi = -\omega t_d$. In the first case the output is a magnified replica of the input, in the second a magnified and inverted replica, and in the third the output is a magnified copy of the input but delayed by a time interval t_d (*see* Problem 1).

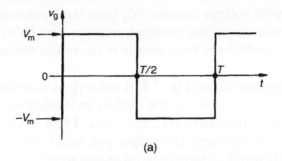

(a)

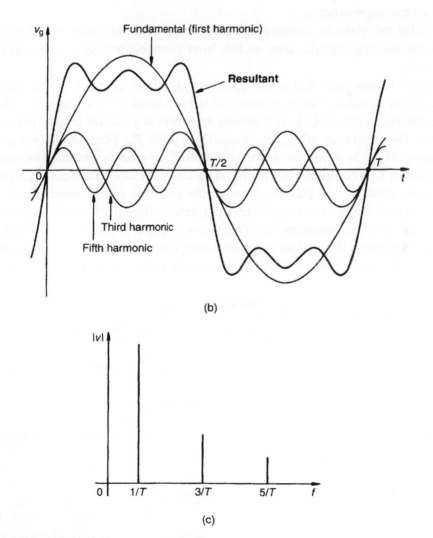

(b)

(c)

Fig. 7.2 (a) An ideal square wave. (b) Showing the first three frequency components of (a) and their sum. (c) Magnitude-frequency spectrum of (b)

7.2   PRACTICAL VOLTAGE AMPLIFIERS

If the components C_c, R_i and C_i, are added to the ideal voltage amplifier of Fig. 7.1(a) then we have a circuit (Fig. 7.3(a)) whose performance approximates that of a practical a.c.-coupled voltage amplifier, C_c being the input coupling capacitor. R_i is the input resistance of the practical amplifier and C_i its effective input capacitance, which results from the presence of strays and device inter-electrode capacitances.

The frequency response plots (Fig. 7.3(b) and (c)), now no longer frequency-independent from $\omega = 0$ to $\omega = \infty$, can usefully be divided into three sections, low, medium and high frequency (l.f., m.f., h.f.). These designations have no absolute meanings but represent the ranges over which the reactances of the capacitances can be ignored, or have to be taken into account, in comparison with other circuit impedances.

Detailed calculations, concerning these comparisons, are made in later sections when considering specific cases so this brief overview merely summarizes those findings.

The m.f. region, regarded as the 'passband', is that frequency range in which the reactance of C_c can be taken as zero and the reactance of C_i as infinite. In the l.f. range the reactance of C_i is still infinite but there is a fall-off in gain-magnitude because the reactance of C_c is comparable with R_i. For the h.f. range, the reactance of C_c is zero but there is a fall-off in gain-magnitude because the reactance of C_i is comparable with R_g. A square waveform is subject to frequency and phase distortion in passing through the amplifier being considered because the l.f. and h.f. frequency components are treated differently from those in the m.f. range: Fig. 7.3(d) emphasizes this. Distortion of any waveform is minimized, in a practical amplifier, if its major frequency components fall well inside the amplifier passband. It is not always necessary to know the precise details of the line spectra

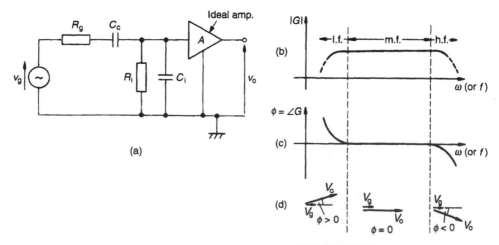

Fig. 7.3   Characteristics of a practical a.c.-coupled voltage amplifier; l.f., m.f., h.f. refer to the low, medium and high frequency ranges

of an input signal to achieve this. Thus, 'fast' edges normally require an m.f. range that extends well into the megahertz region.

In the case of the 'step' edge of an ideal square waveform, there is a simple relationship between acceptable output transition time and the frequency response required to produce it (*see* Section 7.7).

As the boundaries of the m.f. region are dependent on CR-network sinusoidal response, this topic is reviewed next, before an analysis of particular amplifier configurations.

7.3   CR-NETWORK SINUSOIDAL RESPONSE

Figure 7.4 shows a series CR circuit driven by a sinusoidal input voltage of angular frequency ω and RMS magnitude V. There are two popular ways of displaying, graphically, the frequency response of the circuit; the polar plot (an extension of the phasor diagram), and logarithmic frequency plots.

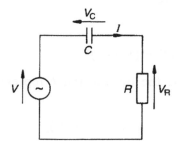

Fig. 7.4   Sinusoidally-driven series CR circuit

The polar plot

Figure 7.5(a) shows a phasor diagram for Fig. 7.4:

$OP = V_R$, $PQ = V_C$, and $OQ = V$

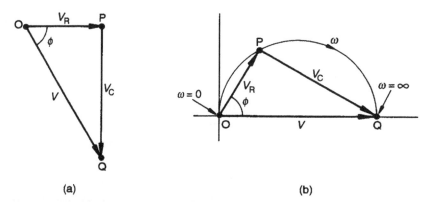

(a)   (b)

Fig. 7.5   Relating to Fig. 7.4. (a) Phasor diagram. (b) Polar plot for V_R

In Fig. 7.4, the maximum power dissipated in R occurs when $\omega \to \infty$ and is given by,

$$P_m = V^2/R \tag{7.17}$$

If P_c = power dissipated in R at $\omega = \omega_c$, then,

$$P_c = (V/\sqrt{2})^2/R \tag{7.18}$$

$$\therefore \quad (P_c/P_m) = 0.5 \tag{7.19}$$

Hence, ω_c is sometimes called a 'half-power' frequency. Equation (7.9) characterizes the frequency response of a simple 'high-pass' filter, ω_c defining what is meant by 'high'. Thus, for $\omega \gg \omega_c$ input signals appear at the output without significant attenuation. Similarly, Equation (7.13) characterizes a simple 'low-pass' filter.

The polar plot shows gain-magnitude and phase on the same diagram and offers physical insight into system transfer functions generally. However, it can be tedious to plot accurately for complex functions.

Logarithmic frequency plots ('Bode' plots)

Plotting gain-magnitude to a linear base of frequency for the audio range (20 Hz $\to$ 20 kHz), alone, requires a piece of graph paper 2 m wide for a horizontal scale 1 mm $\equiv$ 10 Hz. Clearly, a linear frequency scale is unacceptable in practice. However, a logarithmic scale *is* acceptable because it gives range-compression. It also turns out that the use of a logarithmic scale for gain-magnitude, as well as frequency, results in response plots that are well-approximated by piecewise-linear sections for the most frequently encountered transfer functions. Furthermore, plots of phase on a linear scale to a logarithmic base of frequency also yield acceptable piecewise-linear approximations. These piecewise-linear approximations for gain-magnitude and phase are the key to quick and easy analysis and design.

There already exists a logarithmic unit for the ratio (P_2/P_1) of two power dissipations, P_2 and P_1, and that is the 'decibel' (dB). This is 'borrowed' to express voltage ratio in voltage amplifiers. Expressing (P_2/P_1) in dB we have,

$$(P_2/P_1)|_{dB} = 10 \log_{10}(P_2/P_1) \tag{7.20a}$$

When P_2 and P_1 result from RMS voltages V_2 and V_1, occurring across resistors R_2 and R_1, respectively, Equation (7.20a) can be re-written,

$$(P_2/P_1)|_{dB} = 20 \log_{10}|(V_2/V_1)| - 20 \log_{10}(R_1/R_2) \tag{7.20b}$$

In voltage amplifiers it is common practice to use the first term, only, on the right-hand side of Equation (7.20b) to define 'voltage ratio in dB'. Thus,

$$|(V_2/V_1)|_{dB} = 20 \log_{10}|(V_2/V_1)| \tag{7.21}$$

In fact, for any ratio X, it is convenient to write,

$$|X|_{dB} = 20 \log_{10} |X| \tag{7.22}$$

The intended meaning of 'dB' in a particular application is normally clear from the context.

Consider, now, Equation (7.11a). Taking logarithms, to the base 10, of each side and multiplying throughout by 20 gives,

$$|F|_{dB} = 20 \log_{10}[1/\sqrt{1 + (\omega_c/\omega)^2}] \tag{7.23}$$

For $\omega \gg \omega_c$, $\sqrt{1 + (\omega_c/\omega)^2} \simeq 1$

$$\therefore \quad |F|_{dB} = 20 \log_{10} 1 = 0 \tag{7.24}$$

But, for $\omega_c \gg \omega$, $\sqrt{1 + (\omega_c/\omega)^2} \simeq (\omega_c/\omega)$

$$\therefore \quad |F|_{dB} = 20 \log_{10} 1 - 20 \log_{10}(\omega_c/\omega) \tag{7.25a}$$

or,

$$|F|_{dB} = 20 \log_{10} \omega - 20 \log_{10} \omega_c \tag{7.25b}$$

If $y \equiv |F|_{dB}$, $x \equiv \log_{10} \omega$, $c \equiv -20 \log_{10} \omega_c$ then Equation (7.25b) can be written in the form,

$$y = mx + c \tag{7.26}$$

Equation (7.26) describes a straight line with slope m.

With a logarithmic scale a horizontal unit corresponds to a frequency ratio of 10, i.e. a decade. This can be seen from the identity $[\log_{10} 10\omega - \log_{10} \omega] \equiv 1$, ω being any frequency. Hence a comparison of Equations (7.25b) and (7.26) leads to the conclusion that '$m = 20$' means a slope of 20 dB per decade (20 dB/dec).

Sometimes it is more convenient to express the slope in dB/oct, an octave (oct) being a frequency ratio of 2. In this case,

$$20 \log_{10} 2\omega - 20 \log_{10} \omega \equiv 20 \log_{10} 2 = 6.02 \, dB$$

It is normal practice to ignore the 0.02 and say that the slope is 6 dB/oct.

The straight line given by Equation (7.25b) *only* applies for $\omega_c \gg \omega$ but, extrapolated, it passes through $\omega = \omega_c$, for which $|F|_{dB} = 0$.

Equation (7.24) represents a tangent to the true response curve, given by Equation (7.23), and is regarded in this particular case as the 'high frequency asymptote', 'high' meaning $\omega \gg \omega_c$. Similarly, Equation (7.25b) describes a tangent to the true response curve for $\omega \ll \omega_c$ and is regarded as the 'low frequency asymptote'.

Logarithmic frequency plots are usually called 'Bode' plots after their American originator. The description is used to cover both the accurate plots and the piecewise-linear plots of gain-magnitude and phase. We will take it to mean the piecewise-linear versions.

Figure 7.7(a) shows the Bode gain plot for $|F|$, given by Equation (7.11a): the bold sections refer to the piecewise-linear approximation and the faint curve to the accurate plot.

To construct Fig. 7.7(a) a useful procedure is to identify point X at $\omega = \omega_c$ and draw an h.f. asymptote along the 0 dB axis, then measure back to $\omega = (\omega_c/10)$ and down 20 dB to locate point Y. A line through X and Y gives the l.f. asymptote. The error in the piecewise-linear graph is worst at $\omega = \omega_c$: X' on the faint curve is 3 dB (approx.) below X. This can be seen by substituting $\omega = \omega_c$ in Equation (7.23).

Figure 7.7(a) suggests a reason for other adjectives used to describe the frequency ω_c, in addition to those used earlier. These are: break; corner; 3 dB; cut-off; roll-off.

The piecewise-linear phase plot of Fig. 7.7(b) is the result of a curve-fitting procedure. The phase changes by 45°/dec between $(\omega_c/10)$ and 10 ω_c: the error is a maximum ($\sim6°$) at these frequencies but zero at $\phi = 45°$. Figures 7.8(a) and (b)

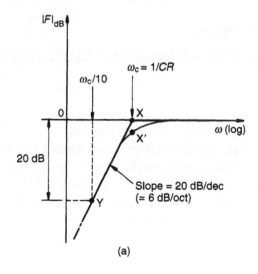

(a)

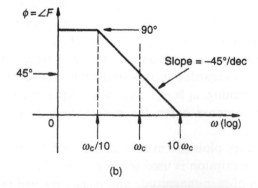

(b)

Fig. 7.7   Bode plots of (V_R/V) for Fig. 7.4. (a) Magnitude. (b) Phase

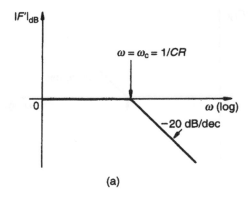

(a)

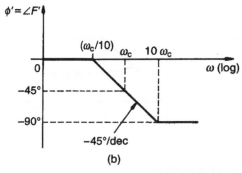

(b)

Fig. 7.8 Bode plots of (V_C/V) for Fig. 7.4. (a) Magnitude. (b) Phase

show $|F'|$, ϕ' given by Equation (7.14). The 'l.f.' asymptote in Fig. 7.8(a) extends to ω_c and the 'h.f.' above that. The graph is obtained by an extension of the algebraic argument used in obtaining Fig. 7.7(a). The phase characteristic is that of Fig.7.7(b) shifted vertically down by 90°.

Example 7.1

Sketch the Bode plots for $F = [1 + j(\omega/\omega_c)]$.

Solution

For $\omega_c \gg \omega$, $F \simeq 1$, $|F|_{dB} \simeq 0$ and $\phi = \angle F = 0°$
For $\omega \gg \omega_c$, $|F|_{dB} = 20 \log_{10}(\omega/\omega_c)$ and $\phi = \tan^{-1}(\omega/\omega_c)$
The Bode plots (Fig. 7.9) are mirror images of those in Fig. 7.8.

Although the function in this example does not occur by itself, it sometimes forms part of a more complex system transfer function. Constant gain, i.e. frequency-independent gain, is characterized by plots parallel to the frequency axis.
  Suppose, now, a transfer function F is of the form,

$$F = F_1 F_2 F_3 \ldots \qquad (7.27a)$$

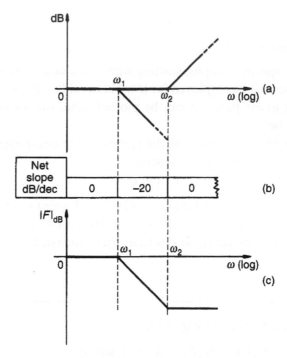

Fig. 7.11 The Bode magnitude-plot, (c), for Fig. 7.10, is constructed from its component parts in (a) and the net-slope table in (b)

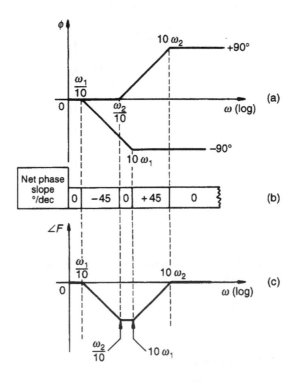

Fig. 7.12 Construction of Bode phase-plot for Fig. 7.10

The construction of the sketches in Figs 7.11 and 7.12 follows the procedural rules given.

(b) From the data supplied, $R_2 = 9 \text{ k}\Omega$ so $(R_1 + R_2) = 10 \text{ k}\Omega$. Hence,

$$\omega_1 = 1/(100 \text{ nF} \times 10 \text{ k}\Omega) = 1000 \text{ rad/s}$$

$$(\omega_1/10) = 100 \text{ rad/s}; \quad 10\,\omega_1 = 10\,000 \text{ rad/s}$$

Also, $\omega_2 = 10\,000$ rad/s; $(\omega_2/10) = 1000$ rad/s; $10\,\omega_2 = 100\,000$ rad/s. From Equation (7.29b), $n \geqslant 3$. In Fig. 7.13, four-cycle paper is used. The shapes of the graphs do not alter if they are plotted to a base of $\log_{10} f$: they are merely shifted to the left along the frequency axis, since $\omega = 2\pi f$ and $\log_{10} f = \log_{10} \omega - \log_{10} 2\pi$.

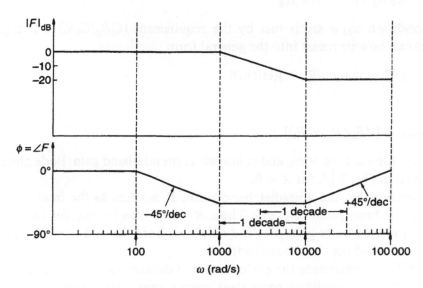

Fig. 7.13 Bode plots for Fig. 7.10 for $R_1 = 1 \text{ k}\Omega$, $R_2 = 9 \text{ k}\Omega$, $C = 100 \text{ nF}$

7.4 AMPLIFIER SYSTEM RESPONSE

Referring to Fig. 7.3(a), let,

$$Z_1 = R_g + (1/j\omega C_c) \tag{7.30a}$$

and,

$$Z_2 = R_i \,\|\, (1/j\omega C_i) = R_i/(1 + j\omega C_i R_i) \tag{7.30b}$$

Then,

$$G = (V_o/V_g) = AZ_2/(Z_1 + Z_2) \tag{7.31a}$$

or,

$$G = A/[1 + (Z_1/Z_2)] \tag{7.31b}$$

Substituting for Z_1, Z_2 the denominator, D, of Equation (7.31b) can be written,

$$D = 1 + (R_g/R_i) + (C_i/C_c) + j\omega C_i R_g + (1/j\omega C_c R_i) \qquad (7.32)$$

$C_c \gg C_i$ for practical amplifiers, so there is insignificant error in approximating the expression for D by the product of three interpretable factors. Thus,

$$D \simeq [1 + (R_g/R_i)][1 - j(\omega_L/\omega)][1 + j(\omega/\omega_H)] \qquad (7.33)$$

The characteristic frequencies ω_L, ω_H define the low and high break points of the frequency response.

$$\omega_L = 1/C_c(R_g + R_i) \qquad (7.34a)$$

$$\omega_H = 1/C_i(R_g \| R_i) = 1/C_i R'_g \qquad (7.34b)$$

The condition $\omega_H \gg \omega_L$ is met by the requirement $(C_i R_g/C_c R_i) \ll 1$. Equation (7.31b) can now be recast into the general form,

$$G = G_{m.f.}/[1 - j(\omega_L/\omega)][1 + j(\omega/\omega_H)] \qquad (7.35)$$

where,

$$G = G_{m.f.} = [AR_i/(R_g + R_i)] \qquad (7.36)$$

$G = G_{m.f.}$ for $\omega_H > \omega > \omega_L$ and is known as the mid-band gain. Bode plots for G are shown in Fig. 7.14: for $A < 0$.

Conventionally, the amplifier bandwidth, B, is taken as the frequency range between the breakpoints ω_L, ω_H. Thus, $B = (\omega_H - \omega_L) \simeq \omega_H$, for $\omega_H \gg \omega_L$. At ω_L, ω_H the true Bode gain plot is 3 dB down from the mid-band gain, so B is sometimes called the '3 dB bandwidth'.

Table 7.1, summarizing the performance of the amplifier scheme in Fig. 7.3(a), indicates that a simplified equivalent circuit, that characterizes the amplifier frequency response, is applicable to each of the three frequency ranges.

Example 7.3 _____

An amplifier has a mid-band gain of -100 and -3 dB points at $f_L = 20$ Hz, $f_H = 20$ kHz.

(a) Write down an algebraic expression for the gain as a function of frequency f.
(b) Calculate the gain-magnitude, in numbers and dB, and the phase at 50 kHz.
(c) Sketch Bode plots.

Solution

(a) In terms of f, expressed in kHz, Equation (7.35) gives,

$$G = -100/[1 - j(0.02/f)][1 + j(f/20)]$$

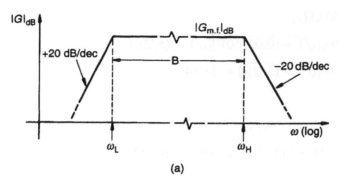

(a)

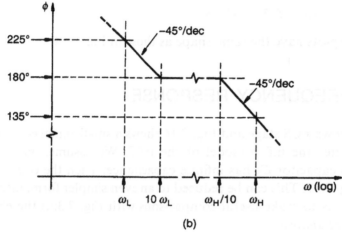

(b)

Fig. 7.14   Bode plots for Equation (7.35) for $A < 0$

Table 7.1

Frequency range	Reactance of C_c	Reactance of C_i	Simplified equivalent circuit
L.F. $\dfrac{\omega_L > \omega}{\omega_L = 1/C_c\,(R_g + R_i)}$	*	∞	
M.F. $\omega_H > \omega > \omega_L$	0	∞	
H.F. $\dfrac{\omega > \omega_H}{\omega_H = 1/C_i\,R'_g}$	0	*	

* Reactance cannot be ignored.

Inequality (7.42) also includes (7.41) because $(C_{ds} + C_{gd}) > C_{gd}$ and $g_{fs} \gg (1/R'_D)$, i.e. the mid-band gain is much greater than unity. Equation (7.43) describes the simplified output circuit. To find the simplified input circuit, substitute V_o from Equation (7.43) into Equation (7.37). Then,

$$I = j\omega[C_{gd}(1 + g_{fs}R'_D)]V = j\omega C_m V \tag{7.44}$$

C_{gd} can thus be removed from the circuit of Fig. 7.17 and its effect taken into account by connecting a larger capacitor, $C_m = C_{gd}(1 + g_{fs}R'_D)$, in parallel with C_{gs}.

Hence, for the condition specified by inequality (7.42), the circuit of Fig. 7.17 can be replaced by the 'unilateral' circuit in Fig. 7.18. (This, in turn, can be replaced by a simplified form shown in Table 7.1 for each of the frequency ranges l.f., m.f., h.f., listed).

Comparing Fig. 7.18 with Fig. 7.3(a), it is apparent that $(C_{gs} + C_m) \equiv C_i$ and $-g_{fs}R'_D \equiv A$. Substituting these values in Equations (7.34) and (7.36) gives the values of ω_L, ω_H, $G_{m.f.}$ to be used for G in Equation (7.35) and Fig. 7.15. Thus,

$$G_{m.f.} = -g_{fs}R'_D R_i/(R_g + R_i) \tag{7.45}$$

$$\omega_L = 1/C_G(R_g + R_i) \tag{7.46}$$

$$\omega_H = 1/(C_{gs} + C_m)(R_g \parallel R_i) \tag{7.47}$$

The discussion leading to Equation (7.44) deals with a particular case of a general phenomenon in electronics.

The magnification in capacitance, owing to an inverting amplifier network connected between the plates of a capacitor, is known as the 'Miller effect' after the American engineer who first noted the effect (in thermionic tubes). In some instances, as here, this magnification is an undesirable feature of circuit operation because it restricts the bandwidth. However, the Miller-effect is also made use of in some circuit designs, particularly when it is necessary to narrow the bandwidth.

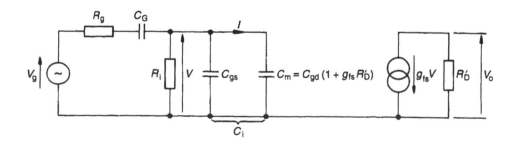

Fig. 7.18   Simplified form of Fig. 7.17, using the Miller approximation

Effect of finite bypass capacitance

In the previous section it was assumed, for simplicity, that $C_S = \infty$. Consequently, ω_L was set by C_G. However, this is not necessarily the case in practice. Figure 7.19 shows that part of Fig. 7.16 that applies at l.f. when $C_S \neq \infty$.

A straightforward, though somewhat tedious, algebraic analysis yields,

$$G_{l.f.} = G_{m.f.}R'_S[1 + j(\omega/\omega_{s1})]/R_S[1 + j(\omega/\omega_{s2})][1 - j(\omega_g/\omega)] \qquad (7.48)$$

$G_{m.f.}$ is given by Equation (7.45). The other quantities in Equation (7.48) are defined as follows,

$$\omega_{s2} = 1/C_S R'_S \qquad (7.49)$$

For the normal case, $r_{ds} \gg R_D$, $\mu(= g_{fs}r_{ds}) \gg 1$,

$$R'_S = R_S \| (1/g_{fs}) \qquad (7.50)$$

Also,

$$\omega_{s1} = 1/C_S R_S \qquad (7.51)$$

and,

$$\omega_g = 1/C_G(R_g + R_i) \qquad (7.52)$$

Figure 7.20 shows a Bode magnitude plot of Equation (7.48) for two cases. In Fig. 7.20(a), an extension of Fig. 7.14(a), $\omega_L = \omega_g \gg \omega_{s2}$: C_S, C_g are chosen so that the effect of C_G dominates. For a specified ω_L the second case requires a smaller C_S, because of the relative magnitudes of $(R_g + R_i)$ and R'_S, so it is particularly relevant in design.

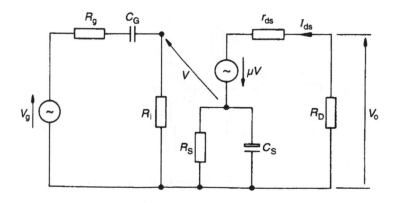

Fig. 7.19   L.F. equivalent of Fig. 7.16 when C_S does not have zero reactance

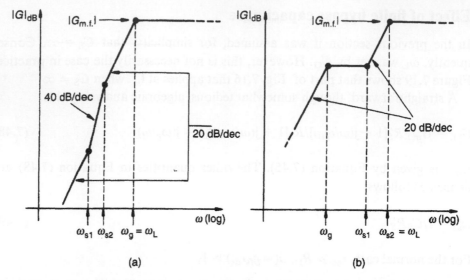

Fig. 7.20   Bode magnitude plots showing the effect of finite C_S. (a) ω_L set by C_G. (b) ω_L set by C_S

7.6   CE FREQUENCY RESPONSE

Figure 7.21 shows a CE amplifier stage and Fig. 7.22 its complete small-signal representation using the hybrid-π model of the BJT. In this, $R_{TH} = (R_1 \| R_2)$ and C_w represents stray wiring capacitance.

As with the CS stage, we can identify three operating ranges, each with its simplified circuit derived from Fig. 7.22. If, in the m.f. equivalent circuit of Fig. 7.23, $R'_C = R_C \| r_o$ and $R_{TH} \gg R_g$ then,

$$G_{m.f.} = -g_m R'_C r_\pi / (R_g + r_x + r_\pi) \tag{7.53}$$

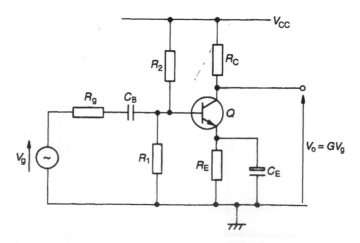

Fig. 7.21   BJT CE amplifier stage

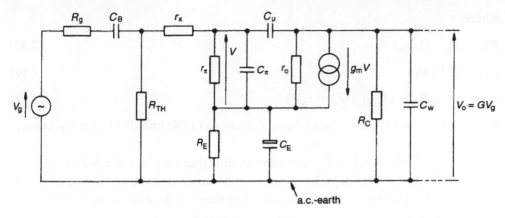

Fig. 7.22 Small-signal equivalent circuit of Fig. 7.21

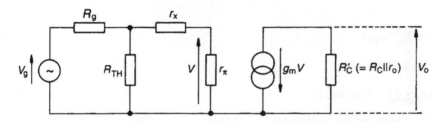

Fig. 7.23 M.F. version of Fig. 7.22

Unfortunately the l.f. analysis of the CS stage is *not* applicable to the l.f. equivalent circuit of the CE stage in Fig. 7.24, because r_π links the input and output loops. However, for order-of-magnitude calculations, break frequencies can be assumed to occur at ω_{e1}, ω_{e2} and ω_h. They are analogous, respectively, to ω_{s1}, ω_{s2} and ω_g, in Section 7.5. Thus,

$$\omega_{e2} = 1/C_E R'_E \qquad (7.54)$$

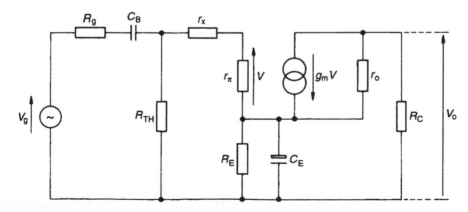

Fig. 7.24 L.F. version of Fig. 7.22

where,

$$R'_E \simeq R_E \| (1/g_m) \tag{7.55}$$

$$\omega_{e1} = 1/C_E R_E \tag{7.56}$$

$$\omega_b = 1/C_B[R_g + \{R_{TH} \| (r_x + r_\pi)\}] \tag{7.57}$$

Values of C_E and C_B calculated from Equations (7.54) and (7.57), for a given ω_L, can be used in a more detailed analysis of circuit operation using a computer. Figure 7.25 shows a CE h.f. equivalent circuit that employs the Miller approximation, discussed earlier. The components inside the dotted contour in (b) constitute the Thévenin equivalent circuit of those in the dotted contour of (a).

For $R_{TH} \gg R_g$,

$$V'_g \simeq V_g r_\pi/(R_g + r_x + r_\pi) \tag{7.58}$$

and,

$$R_b \simeq r_\pi \| (R_g + r_x) \tag{7.59}$$

Hence, the h.f. gain is,

$$G_{h.f.} = G_{m.f.}/[1 + j(\omega/\omega_H)] \tag{7.60}$$

where, $G_{m.f.}$ is given by Equation (7.53) and,

$$\omega_H = 1/C_i R_b = 1/(C_\pi + C_m)R_b \tag{7.61a}$$

or,

$$\omega_H = 1/[C_\pi + C_\mu(1 + g_m R'_C)]R_b \tag{7.61b}$$

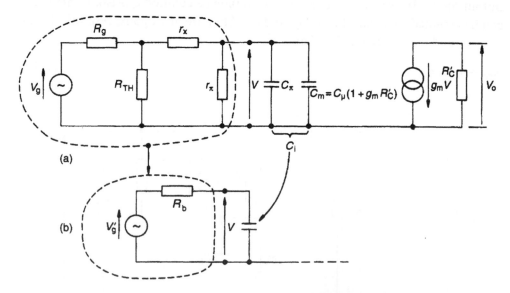

Fig. 7.25   (a) H.F. version of Fig. 7.22, using the Miller approximation. (b) Simplified form of input circuit for (a)

From the reasoning in Section 7.5, ω_H, as given by Equation (7.61b), is applicable provided,

$$\omega_H \ll 1/(C_\mu + C_w)R'_C \tag{7.62}$$

Calculation of ω_H in Equation (7.61b) requires a knowledge of C_μ and C_π. Manufacturers normally quote C_{ob} ($\simeq C_\mu$) on data sheets. It is independent of I_{CQ} but has a voltage (V_{CBQ}) dependence, characteristic of a reverse-biased PN junction.

C_π ($\gg C_\mu$) is not given on data sheets. It must be deduced from a specified parameter, ω_T, that characterizes the frequency dependence of CE current gain, as will be evident from a consideration of Fig. 7.26. In Fig. 7.26(a), I_{BQ} is the base bias current and I_b is a superimposed small-signal variation that causes a change I_c in the collector bias current I_{CQ}. The small-signal equivalent circuit is shown in Fig. 7.26(b).

Applying KCL at B' and C gives,

$$I_b = V[(1/r_\pi) + j\omega C_\pi] + I \tag{7.63}$$

$$I = j\omega C_\mu V \tag{7.64}$$

$$I_c = g_m V - I \tag{7.65}$$

From Equations (7.63), (7.64) and (7.65) it follows that

$$\beta(j\omega) = (I_c/I_b) = (g_m - j\omega C_\mu)/[(1/r_\pi) + j\omega(C_\pi + C_\mu)] \tag{7.66}$$

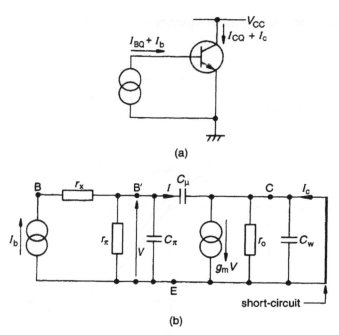

(a)

(b)

Fig. 7.26 Circuits used to define $\beta(j\omega)$. (a) Showing total currents. (b) Small-signal equivalent circuit

For $\omega \ll g_m/C_\mu$,

$$\beta(j\omega) \simeq g_m r_\pi/[1 + j\omega(C_\pi + C_\mu)r_\pi] \tag{7.67}$$

or,

$$\beta(j\omega) = \beta(0)/[1 + j(\omega/\omega_\beta)] \tag{7.68}$$

where,

$$\beta(0) = g_m r_\pi \tag{7.69}$$

and,

$$\omega_\beta = 1/(C_\pi + C_\mu)r_\pi \tag{7.70}$$

The $j\omega$ in brackets after β is used to emphasize the dependence of current gain on frequency, it being understood that $V_{CBQ} > 0$. ($\beta_0(j\omega)$ would refer to the abnormal d.c. operating condition $V_{CBQ} = 0$.) With this notation, $\beta(0)$ is the symbol that should be used for CE current gain in l.f. calculations. However, it is common practice to use the symbol β by itself (as in Chapters 3 and 4) for d.c. and l.f. circuit analyses, when there is no possibility of confusion.

A Bode plot for $|\beta(j\omega)|$ is shown in Fig. 7.27.

For $\omega \gg \omega_\beta$, Equation (7.68) gives,

$$\omega|\beta(j\omega)| = \beta(0)\omega_\beta \tag{7.71}$$

At the 'transition', or 'unity-gain' frequency, $\omega = \omega_T$, $|\beta(j\omega)| = 1$.

$$\therefore \quad \omega_T = \beta(0)\omega_\beta \tag{7.72}$$

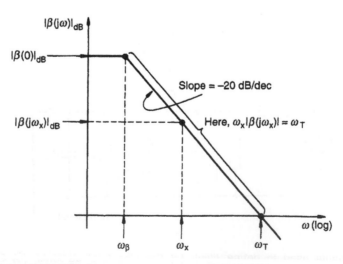

Fig. 7.27   Bode plot for $|\beta(j\omega)|$

Substituting for $\beta(0)$ from Equation (7.69) and ω_β from Equation (7.70), Equation (7.71) gives,

$$\omega_T = g_m/(C_\pi + C_\mu) \tag{7.73}$$

Hence,

$$C_\pi = (g_m/\omega_T) - C_\mu \tag{7.74}$$

ω_T is usually specified at a given I_{CQ}, so C_π can be calculated.

ω_T is sometimes called the 'gain-bandwidth product'. For $\omega_T > \omega_x \gg \omega_\beta$, $\omega_x|\beta(j\omega_x)| = \omega_T$. Hence ω_T can be inferred from a measurement of $|\beta_x(j\omega)|$ at $\omega_x \ll \omega_T$.

7.7 SAG, RISE-TIME AND BANDWIDTH

The response of an amplifier to a small voltage step provides a guide to the bandwidth required to preserve the shape of an arbitrary waveform during the amplification process. ('Small' has the meaning it was given in earlier chapters in discussing the applicability of equivalent circuits.)

Table 7.2 is a pictorial summary of the operation of the circuit in Fig. 7.3(a) when v_g is a step of amplitude V applied at $t = 0$.

The response can be divided into two consecutive sections. During the edge C_c behaves as a short-circuit so the equivalent circuit is that in 1(a). As shown in 2(a), the output reaches a peak at $5C_iR_g'$ (approximately). The 10–90% rise time, t_r, can be calculated from $v_o(t)$, quoted in 3(a). (CR step response is covered

Table 7.2

	Small signal step response		
	1	2	3
	Applicable circuit	Output waveshape	$v_o = f(t)$
(a)	$R_g' = R_g \| R_i$	Rise-time	$t \leqslant 5C_iR_g'$ $v_o = Av_i$ $\quad = G_{m.f.}V[1 - e^{-t/C_iR_g'}]$ $t_r \simeq 2.2C_iR_g' \simeq 2.2/\omega_H$ $t_r \simeq 0.35/f_H$
(b)			$t' \gg 5C_iR_g'$ $v_o = Av_i$ $\quad = G_{m.f.}Ve^{-t'/C_c(R_g + R_i)}$ % sag at $t' \simeq t' \times 200\pi f_L$ provided $t' \ll C_c(R_g + R_i)$

in Chapter 10). It is related to the 3 dB point f_H by,

$$t_r = 0.35/f_H \simeq 0.35/B \text{ (Hz)} \tag{7.75}$$

Thus, a step applied to the Y amplifier of an oscilloscope with a bandwidth of 350 MHz produces an output pulse with a 10 ns rise-time.

For $t > 5C_i R'_g$, the equivalent circuit is shown in 1(b). This corresponds to the discharge of C_c, during which the current in C_i is negligible. The discharge process causes a 'sag' or 'tilt' on the output waveform in 2(b). The magnitude of this sag can be calculated from the discharge equation, quoted in 3(b), and related to the 3 dB freqency f_L.

$$\text{Percentage sag} = [\text{sag}/v_o(\text{max})] \times 100 \tag{7.76}$$

For $C_c(R_g + R_i) \gg t' > 5C_i R'_g$

$$\% \text{ sag} = 100[t'/C_c(R_g + R_i)] \simeq (t' \times 200\pi f_L) \tag{7.77}$$

For a square wave of frequency $f (= 1/T)$,

$$\% \text{ sag(max)} \simeq 100\pi f_L/f. \tag{7.78}$$

7.8   SELF ASSESSMENT TEST

1 What is meant by the frequency spectrum of a waveform?

2 What are the conditions for distortionless amplification?

3 What is the gain in decibels of a voltage amplifier having a numerical gain of -10^3?

4 What is a Bode plot?

5 Define the parameters ω_β and ω_T of a BJT.

6 What is meant by the 'Miller Effect'?

7 How is the bandwidth of an oscilloscope Y amplifier related to the rise-time?

7.9   PROBLEMS

1   The gain-magnitude, A_o, of an amplifier is frequency-independent but the phase shift, ϕ, is frequency-dependent and given by $\phi = -\omega t_d$. The input is a square wave of peak value V and periodic time T. Show that the output due to the third harmonic component of the input waveform is

$(4VA_o/3\pi) \sin 3\omega_1(t - t_d)$ where, $\omega_1 = 2\pi/T$.

Hence, show that the output is a magnified replica of the input, delayed by a time t_d.

2   A two-stage d.c.-coupled amplifier has a gain function of the form,

$$A = 100/[1 + j(f/f_H)]^2$$

where, $f_H = 0.5$ MHz.

(a) Write down expressions for $|A|$, $\angle A$ at any frequency, f.
(b) Calculate $\angle A$ for $|A| = 50$.
(c) Draw a polar plot of A, based on calculations of $|A|$, $\angle A$ at representative frequencies in the range $\infty \geqslant f \geqslant 0$.

3   Show that, compared with the true Bode plots, the piecewise-linear gain plots of Fig. 7.8(a) and 7.9(a) are in error by 1 dB (approx.) at $\omega = (\omega_c/2)$ and $\omega = 2\omega_c$.

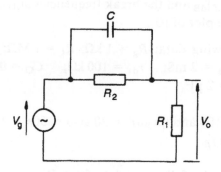

Figure 7.28

4   Figure 7.28 shows a phase-advance network.

(a) Show that $F = (V_o/V_g) = K[1 + j(\omega/\omega_1)]/[1 + j(\omega/\omega_2)]$, where: $K = R_1/(R_1 + R_2)$; $\omega_2 = 1/C(R_1 \parallel R_2)$; $\omega_1 = 1/CR_2$.
(b) Show that the Bode plots of $|F|$, $\angle F$ have shapes that resemble mirror images (in the frequency axis) of the plots shown, respectively, in Figs 7.11(c) and 7.12(c).

5   Establish the justification for Equation (7.33), by multiplying the bracketed terms and assuming $C_cR_g \gg C_iR_g$.

6   An amplifier has -3 dB points at 100 Hz and 100 kHz and a mid-band gain of -50.

(a) Write down an algebraic expression for the gain A as a function of frequency f.

(b) Calculate $|A|$, $\angle A$ at $f = 25$ Hz and $f = 300$ kHz.
(c) Sketch Bode plots for $|A|$, $\angle A$.

7   A d.c.-coupled amplifier is constructed from n identical stages connected in cascade. Each stage has an infinite input impedance and zero output impedance and a gain function of the form,

$$A = A_o/[1 + j(f/f_H)]$$

(a) Show that the composite gain-magnitude, $|A_c|$, at any frequency, f, is given by,

$$|A_c| = A_o^n/[1 + (f/f_H)^2]^{n/2}$$

(b) Show that the overall (-3 dB) cut-off frequency, f_H^*, is given by $f_H^* = rf_H$ where, r = bandwidth reduction factor = $\sqrt{(2^{1/n}) - 1}$
(Hint: $f = f_H^*$ when $|A_c| = A_o^n/\sqrt{2}$.)

8   For the CS circuit of Fig. 7.15,

(a) Calculate $|G_{m.f.}|_{dB}$ and the break frequencies ω_{s1}, ω_{s2}, ω_g and ω_H.
(b) Sketch a Bode plot of $|G|$.

Assume the following data: $R_g = 1$ kΩ; $R_1 = 1$ MΩ; $R_2 = \infty$; $R_S = 1$ kΩ; $R_D = 10$ kΩ; $g_{fs} = 2$ mS; $r_{ds} = 100$ kΩ; $C_G = 0.1$ μF; $C_S = 50$ μF; $C_{gs} = 5$ pF; $C_{gd} = 2$ pF.

9   A BJT has $\beta(0) = 100$ and $|\beta(j\omega)| = 20$ at $\omega = 50\pi \times 10^6$ rad/s.
Calculate f_T and f_β.

10   A data sheet gives the following data for a BJT: $r_x = 100$ Ω; $V_A = 100$ V; $\beta(0) = 100$ at $V_{CB} = 7.5$ V; $C_\mu = 2$ pF and $f_T = 500$ MHz at $V_{CB} = 7.5$ V, $I_C = 2.5$ mA. Calculate the parameters r_π, C_π and r_o of the hybrid-π model.

11   The BJT specified in problem 10 is operated in the circuit of Fig. 7.21, under the same d.c. conditions. Component data are as follows: $R_g = 100$ Ω; $R_C = 2.7$ kΩ; $R_E = 390$ Ω; $(R_1 \| R_2) \gg 100$ Ω; C_E and C_B are large enough to give $\omega_L \simeq 100$ Hz. Calculate f_H.

12   A 10 kHz square wave is a.c.-coupled to the Y input channel of an oscilloscope. Calculate f_L and f_H of the Y amplifier if the waveform observed on the oscilloscope screen exhibits a 10% sag (max.) and a 10–90% rise-time of 20 ns.

8 Feedback

Feedback exists in an amplifier system when a part, or the whole, of an output signal appears in the input circuit. This can occur unintentionally, with effects that can be detrimental to the amplification process, or intentionally with the object of obtaining certain desired modifications to the amplifier performance.

Unintentional feedback may exist because of common coupling effects in the supply rail leads and this generally produces undesirable results. An everyday example of the effect of unintended feedback is the 'howl' in a public address system when some of the acoustic energy produced by a loudspeaker finds its way back to a live microphone via reflections from the walls of an enclosure.

Intentional feedback is applied to improve one, or more, of the characteristics of an amplifier, e.g. the voltage gain magnitude, $|A|$, which is often poorly defined. In the case of a single stage CE BJT amplifier, driven from a low resistance signal source, $|A| \simeq g_m R_C$ for audio frequency signals (*see* Chapter 4). For base-potentiometer bias, the I_C tolerance, on which the magnitude of g_m depends, generally exceeds $\pm 10\%$ because of supply rail, base emitter voltage and emitter resistor tolerances. Allowing for the tolerance on R_C and the effect of possible ambient temperature variations, the tolerance on $|A|$ is unlikely to be less than $\pm 15\%$. With two or more stages in cascade, for increased gain, the overall gain uncertainty is worse. FET stages offer no more certainty.

This presents a problem. In instrumentation and measurement systems, for example, gain tolerances of $\pm 2\%$ are often required. The problem can be solved by the appropriate application of feedback. The use of feedback is so important that it is difficult to envisage analogue electronics without it.

Feedback exists, of course, in the field of human activity, often with the person involved being unaware of it. Examples are in the process of threading a piece of cotton through the eye of a needle and in driving an automobile along a busy motorway. In each case, the eye continuously senses distance and direction information and sends signals to the brain, which controls the movement of the limbs in order to provide corrective action.

This chapter deals quantitatively with the benefits that arise through the appropriate use of feedback. For this introductory treatment, the amplifier is regarded primarily as a black box. In practice, it can often be treated like this, as will be seen when the subject of feedback is explored in more detail in the next chapter ('Operational Amplifiers'). However, some attention is paid, also, to the circuit implementation of feedback schemes.

8.1   THE FEEDBACK EQUATION

Figure 8.1(a) shows the block diagram of a voltage amplifier with feedback (fb). Amplifier A and network β are assumed to be unilateral non-interacting blocks. Thus, signal flow in A, which is assumed to have an infinite input impedance and zero output impedance, is from left-to-right only, as indicated by the arrow inside the box. Similarly, signal flow is from right-to-left, only, in β which is also assumed to have infinite input impedance and zero output impedance. The gain, A ($= V_o/V_i$), may be a real number, positive ($+A_o$) or negative ($-A_o$), or it may be complex, e.g. $A = -A_o/[1 + j\,(f/f_H)]$, indicating a frequency-dependent gain-magnitude and phase shift. Additionally, β may be real or complex. Typically, however, it is a positive real fraction, being determined by a resistive potentiometer network.

Although A and β may represent four-terminal networks they are, for simplicity, assumed to be three-terminal networks in the small-signal equivalent circuit of Fig. 8.1(b), i.e. one terminal is common to the input and output of each block. The feedback voltage is V_{fb} ($= \beta V_o$).

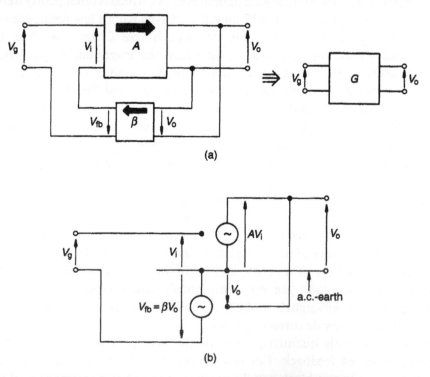

(a)

(b)

Fig. 8.1   A feedback (fb) amplifier configuration. (a) Block schematic (b) Idealized equivalent circuit

Taking due account of the phasor nature of the voltages involved, it follows from Fig. 8.1 that,

$$V_i = V_g + V_{fb} = V_g + \beta V_o \qquad (8.1)$$

But,

$$V_i = V_o/A \qquad (8.2)$$

Hence,

$$(V_o/A) = V_g + \beta V_o \qquad (8.3)$$

$$\therefore G = (V_o/V_g) = A/(1 - A\beta) \qquad (8.4a)$$

Equation (8.4a) is the classic form of the fb equation, derived originally (with alternative symbols) by an American engineer, Black (a flash of inspiration while aboard a ferryboat, on his way to work at the Bell Laboratories).

Definition of terms

A and β are connected in a simple loop and that fact explains some of the nomenclature associated with fb amplifiers. A (or A_v) is the gain of the amplifier without fb. It is also called the 'forward-gain'. β is the 'feedback fraction' or 'feedback ratio'. It is sometimes written B, or b, in order to avoid confusion with the CE current gain of a BJT. (No confusion should arise in a discussion of fb at a system level.)

G is the gain of the amplifier with fb, or the 'closed-loop' gain. (Note the position of the hyphen.) It is also written A', A_{CL}, or A_{VCL}, but G is preferred here, to avoid the use of primes or subscripts.

$A\beta$ is the 'loop-gain' (LG). It is the overall gain of the forward amplifier and fb network connected in cascade. It is also known as the 'loop-amplification'. Figure 8.2 emphasizes the loop meaning of $A\beta$. V_g is set to zero and the loop is cut. A test signal, V_x, applied at the amplifier input side of the cut, produces a voltage, V'_x, on the other side: then $(V'_x/V_x) = A\beta$. In Fig. 8.2 the cut is made right at the input terminals of A, but the same result would have been achieved had a cut been made

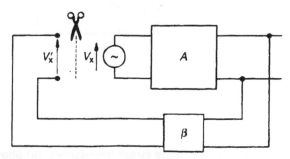

Fig. 8.2   Loop-gain (LG) definition for Fig. 8.1: LG $= (V'_x/V_x) = A\beta$

at any other convenient point in the loop: the loop-gain is a property of the loop alone. In words, Equation (8.4a) reads,

Gain with fb = Gain without fb/[1 − (loop-gain)]   (8.4b)

$(1 − A\beta)$ is the 'feedback factor' and is given the symbol F. It characterizes the 'amount of fb': $F_{dB} = 20 \log_{10} |1 − A\beta|$. A less common description of $(1 − A\beta)$ is 'return-difference': referring to Fig. 8.2, $(1 − A\beta)$ is the difference between V_x and the (returned) voltage V'_x when $V_x = 1$ V.

Impedance effects

Equation (8.4a) was derived with the assumption that A and β were ideal blocks. In practice A and β have impedances associated with them. It is shown here that the form of the classic fb equation is still valid provided a suitable interpretation is given to the meaning of forward-gain and loop-gain.

Figure 8.3(a) is a modified version of Fig. 8.1(b). The symbols used have the following meanings: Z_g is the signal source impedance; Z_{iA}, μ, Z_{oA} are

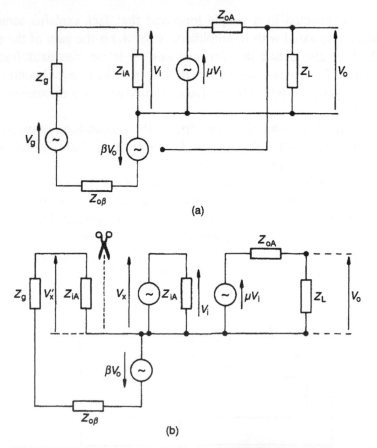

(a)

(b)

Fig. 8.3   Loop-gain interpretation with finite circuit impedances. (a) Showing loop impedances. (b) Procedure on opening the loop

respectively, the input impedance, open circuit voltage gain and output impedance of block A. (The input impedance, $Z_{i\beta}$, of the β block is taken as infinite: in practice, the loading effect of the β network on the A block is often quite small.) In the input circuit the net drive voltage is $(V_g + \beta V_o)$.

From the potential divider rule,

$$V_i = (V_g + \beta V_o)Z_{iA}/(Z_{iA} + Z_g + Z_{o\beta}) \tag{8.5}$$

For the output circuit,

$$V_o = \mu V_i Z_L/(Z_{oA} + Z_L) \tag{8.6}$$

Eliminating V_i from the Equations (8.5) and (8.6) gives,

$$V_g + \beta V_o = V_o/A \tag{8.7}$$

where,

$$A = [\mu Z_L/(Z_L + Z_{oA})] \times [Z_{iA}/(Z_{iA} + Z_g + Z_{o\beta})] \tag{8.8}$$

Equation (8.7) brings us back to Equation (8.4a).

Thus, provided the forward-gain has the definition given in Equation (8.8), the form of the classic fb equation is unaltered when finite circuit impedances are considered. Consequently, we will ignore these impedances in block diagram analyses, except when it is necessary to calculate the effect that fb has on them. The loop-gain is, as before, $A\beta$.

From Equation (8.8) we can make a formal deduction that supports intuitive reasoning, namely, that in making a cut in the fb loop (a 'thought' experiment or an experimental test) to determine the loop-gain, the cut must be terminated by the impedance that existed there before the cut was made. Figure 8.3(b) illustrates this important point for the case where a cut is made near the amplifier input: $(V'_x/V_x) = A\beta$, A having the value given in Equation (8.8).

Types of feedback

Rearranging Equation (8.4a) and taking magnitudes,

$$|G|/|A| = 1/|(1 - A\beta)| \tag{8.9}$$

There are four cases to consider:

(a) $|(1 - A\beta)| > 1$, corresponding to $|G| < |A|$.

This is 'negative', 'inverse' or 'degenerative' fb (nfb)

(b) $|(1 - A\beta)| < 1$, corresponding to $|G| > |A|$.

This is 'positive' or 'regenerative' fb (pfb)

(c) $|(1 - A\beta)| = 1$, i.e. $|G| = |A|$, corresponding to 'neutral' fb.
(d) $|(1 - A\beta)| = 0$.

From Equations (8.1) and (8.2), $V_g = V_i(1 - A\beta)$. It follows that V_i can be finite with $V_g = 0$, if $(1 - A\beta) = 0$, and an output voltage can exist without an externally applied input voltage.

Condition (a) is a design choice for amplifiers; (c) is not a design objective but may occur as a by-product of (a); (d) is carefully avoided in amplifier design but exploited in oscillator circuits.

This chapter is principally concerned with the advantages associated with (a).

Alternative view of feedback

The formulation of Equation (8.4a) followed a convention popular among electronics engineers, that the signal fed back is added algebraically to the externally applied signal. The analysis makes no assumptions about the polarity of the fb. As we have seen above, this is defined according to the magnitude of $(1 - A\beta)$ and that is dependent on signs subsequently attached to A and β.

However, a different approach is normally adopted by control-systems engineers involved in the design of error-actuated systems. It is illustrated in Fig. 8.4. The fb signal is intended to be *subtracted* from the applied signal by some mixing process in the input circuit because *negative* fb is assumed right from the start. The difference between the two conventions leads to a '+' sign instead of a '−' sign in the denominator of Equation (8.4a), and elsewhere, for the product $A\beta$. As will be seen later, a '+' sign also occurs for nfb in the convention adopted here but that is *after A* and β have been allocated signs.

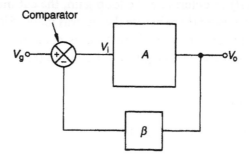

Fig. 8.4   Control engineers' representation of fb connection

The two different conventions arose because fb amplifier theory and control theory developed along separate, although parallel, paths. Provided the reader adheres rigidly to one convention, sign problems are unlikely to arise.

Benefits of negative feedback

Negative feedback is widely used in the design of amplifiers, and electronic equipment generally, because of the host of benefits it brings. These are listed below, with typical application areas. They refer specifically to the configuration of Fig. 8.1. The justification for the benefits is dealt with in subsequent sections.

- Reduction in gain-uncertainty, resulting from component tolerances, supply rail variations and temperature changes, etc. (precise instrumentation).
- Reduction in non-linear, frequency and phase distortion (audio high-fidelity equipment).
- Increase in input impedance (electrometer amplifiers for measuring very low voltages and currents).
- Reduction in output impedance (power supplies).
- Extension in bandwidth (wideband video amplifiers for oscilloscopes and TV).

8.2   NFB: 'A', β FREQUENCY-INDEPENDENT

To illustrate the meaning and application of Equation (8.4a) we assume, in this section, that A and β are both real numbers. Thus, A can be $+A_o$ or $-A_o$, A_o being the gain-magnitude. Similarly, β can be $+\beta_o$ or $-\beta_o$, β_o being the magnitude of the fb ratio. Table 8.1 shows the four possible combinations. Taking due account of the signs of the (real) quantities A and β the fb polarity is given by the sign of their product. Thus, nfb corresponds to $A\beta < 0$, as shown in the middle two columns. The phasors for these are shown in Fig. 8.5, (a) corresponding to column 3 and (b) to column 4. Figure 8.6 shows alternative input circuit representations of the voltage fed back.

Table 8.1   Feedback options when A and β are both real numbers

A	$+A_o$	$+A_o$	$-A_o$	$-A_o$
β	$+\beta_o$	$-\beta_o$	$+\beta_o$	$-\beta_o$
$\angle A\beta$	$0°$	$180°$	$180°$	$0°$
Sign of $A\beta$	$+$	$-$	$-$	$+$
$|G|$	$>|A|$	$<|A|$	$<|A|$	$>|A|$
Fb type	pfb	nfb	nfb	pfb

The denominator of Equation (8.4a) can *now* be written $(1 + A_0\beta_0)$. Hence,

$$G = \pm A_o/(1 + A_o\beta_o) \qquad (8.10a)$$

or,

$$1/|G| = (1/A_o) + \beta_o \qquad (8.10b)$$

The sign in the numerator of Equation (8.10a) depends on the sign of A. A graphical construction for finding $|G|$ is shown in Fig. 8.7, in which the diagram lettering is chosen to avoid confusion with amplifier parameters. The construction is based on Equation (8.10b) (*see* Problem 3). The procedure is as follows.

- Draw a horizontal line HJ of arbitrary length.

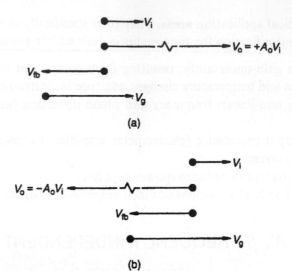

(a)

(b)

Fig. 8.5 Phasors for Table 8.1. (a) Corresponds to column 3. (b) Corresponds to column 4

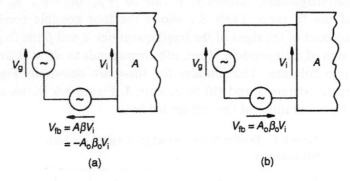

(a)

(b)

Fig. 8.6 Input circuit representations of nfb, for Table 8.1

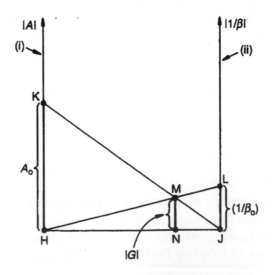

Fig. 8.7 Graphical construction for $|G|$ when A and β are real numbers

- Erect perpendicular lines, (i), (ii) at H, J respectively. These lines are for the display, to scale, of A_o, (i.e. $|A|$) and $1/\beta_o$, (i.e. $|1/\beta|$) respectively.
- Mark point K on line (i), making KH $\equiv A_o$, and mark point L on line (ii) so that LJ $\equiv (1/\beta_o)$.
- Construct the diagonals KJ, LH and mark the point of intersection, M.
- Draw a vertical line through M and mark the point of intersection, N, with HJ.
- MN = $|G|$. Note that the limit value of $|G|$ as $A_o \to \infty$ is $(1/\beta_o)$, i.e. M becomes coincident with L, and N with J.

Example 8.1

An amplifier has $A = -400$ and $\beta = +0.0475$.

(a) Calculate G using Equation (8.4a).
(b) Draw a phasor diagram for an RMS output voltage of 200 mV.
(c) Find $|G|$ by the graphical construction described above.

Solution

(a) $G = -400/[1 - (-400 \times 0.0475)] = -20$
(b) RMS voltage at amplifier input terminals = $(200/400)$ mV = 0.5 mV
    RMS voltage fed back = (200×0.0475) = 9.5 mV
    RMS generator voltage = $(9.5\,\text{mV} + 0.5\,\text{mV})$ = 10 mV (*see* Fig. 8.8).

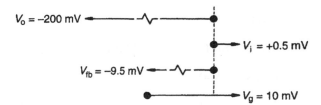

Fig. 8.8   Phasor plot for Example 8.1

(c) For a graphical construction, useful guidelines are: HJ $\equiv$ 16 cm; vertical scale, 1 cm $\equiv$ 25 gain units.

The reduction in gain-uncertainty that is achieved with nfb can be appreciated, pictorially, from Fig. 8.9. Suppose there is a change ΔA_o, of arbitrary magnitude, in A_o. This is shown as an increase but the discussion is equally applicable to a decrease.

A proportionate change in $|G|$ would be $|\Delta G'|$ represented by the line M'M$_0$. However, the actual change is $|\Delta G|$.

As $|\Delta G| < |\Delta G'|$, it follows that $|\Delta G/G| < (\Delta A_o/A_o)$.

To quantify the change in G for the *general* case of fb, we proceed as follows. From Equation (8.4a), rearranged,

$$(1/G) = (1/A) - \beta \tag{8.11}$$

To find G these value are substituted in Equation (8.4a), as is shown in the example that follows.

Despite the particular forms assumed for A and β the approach adopted in this section is applicable in the general case where A and β have arbitrary frequency-dependencies.

Example 8.3

$A = -100/[1 + j(f/f_{\mathrm{H}})]$, where $f = 1$ MHz and $\beta = +0.03$. Calculate $|G|$, $\angle G$ at $f = 0$ MHz and $f = 0.5$ MHz.

Solution

At $f = 0$: $A = -100$; $G = -100/[1 - (-0.03 \times 100)] = -25$

$|G| = 25$, $\angle G = 180°$

At $f = 0.5$ MHz, $A = -100/(1 + j0.5)$

$G = [-100/(1 + j0.5)]/[1 + \{3/(1 + j0.5)\}]$

i.e.

$G = -100/(4 + j0.5) = -25/(1 + j0.125)$

Hence,

$|G| = 25\sqrt{1^2 + (0.125)^2} = 24.8$

and,

$\angle G = 180° - \tan^{-1}(0.125) = 172.87°$

Figure 8.10 shows phasor relationships for the voltages in Fig. 8.1 when A and β have the forms given in Equations (8.18) and (8.19). V_{i} is chosen as the reference phasor and is represented by OP. Semicircle (i), with centre at $-A_{\mathrm{o}}V_{\mathrm{i}}/2$ on the

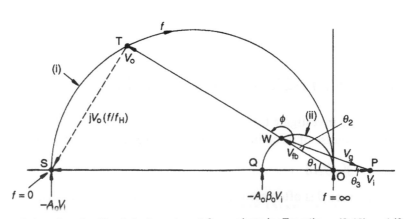

Fig. 8.10   Polar plots for Fig. 8.1 when A and β are given by Equations (8.18) and (8.19)

horizontal axis and radius $A_o V_i/2$, is a polar plot of V_o, i.e. it shows the locus of the tip of the phasor for V_o as f varies from 0 to ∞ with V_i fixed. This follows from a rewritten form for Equation (8.18),

$$V_o + jV_o(f/f_H) = -A_o V_i \tag{8.20}$$

Similarly, semicircle (ii), with centre at $-A_o \beta_o V_i/2$ on the horizontal axis and radius $A_o \beta_o V_i/2$ is a polar plot of V_{fb}. At an arbitrary frequency: $OT \equiv V_o$; $OW \equiv \beta V_o = V_{fb}$; $WP \equiv V_g$.

Without nfb: $|G| = |OT|/|OP|$; $\angle G = 180° - \theta_1$
With nfb: $|G| = |OT|/|WP| = |OW|/(\beta_o|WP|)$; $\angle G = \phi = 180° - \theta_2$

But, by geometry, $\theta_1 = (\theta_2 + \theta_3)$ so $\theta_2 < \theta_1$ and the phase shift introduced by the frequency-dependence of A is reduced by the effect of nfb.

It is normal practice on polar plots of nfb amplifiers to make $OP = 1$ unit. Then: $OS \equiv -A_o$; $OT \equiv A$; $OW \equiv A\beta$; $WP \equiv (1 - A\beta)$. The graph of $A\beta$ is then called a Nyquist plot after the engineer who made pioneering contributions to the theory of feedback. The sections below correspond to the benefits listed in Section 8.1. The general case of fb is considered but, in interpreting the results, it must be remembered that for nfb, $|1 - A\beta| > 1$.

Reduction in gain-uncertainty

The derivation of Equation (8.15) does not depend on A and β being real numbers so we can use it when A is a complex number, as in the case under discussion.

Reduction in distortion

We consider here only non-linear distortion. Frequency and phase distortion are discussed later in 'Extension in bandwidth'.

A non-linear relationship between V_o and V_i, for A, leads to a distorted output signal. This effect is taken into account by the voltage generator V_d in Fig. 8.11. In the absence of the fb (Fig. 8.11(a)),

$$V_{o1} = AV_{i1} + V_d \tag{8.21}$$

With fb (Fig. 8.11(b)),

$$V_{o2} = AV_{i2} + V_d \tag{8.22}$$

and,

$$V_{i2} = V_g + \beta V_{o2} \tag{8.23}$$

From Equations (8.22) and (8.23),

$$V_{o2} = [AV_g/(1 - A\beta)] + [V_d/(1 - A\beta)] \tag{8.24}$$

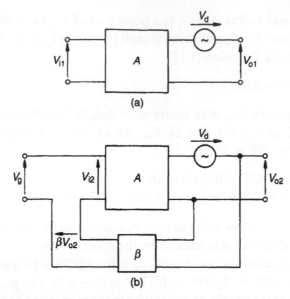

Fig. 8.11   (a) V_d models distortion, due to A, without fb. (b) Equivalent circuit with fb applied

Equation (8.24) shows that the signal and distortion are changed by the same factor $1/(1 - A\beta)$ when fb is applied. Suppose, however, that V_g is adjusted so that $V_g/(1 - A\beta) = V_{i1}$. Then,

$$V_{o2} = AV_{i1} + [V_d/(1 - A\beta)] \qquad (8.25)$$

Comparing Equations (8.21) and (8.25) it is apparent that the effective magnitude of the distortion at the output is reduced by a factor $(1 - A\beta)$ provided the input signal level is increased by the same factor. The success of this scheme depends on the use of a pre-amplifier that does not introduce significant distortion. Such an arrangement is used in the design of audio power amplifiers in which the output stage, operating at high signal levels, normally exhibits non-linear behaviour.

Increase in input impedance

Figure 8.12 shows the equivalent input circuit with fb, Z_{iA} being, as in Section 8.1, the input impedance of A. By inspection,

$$V_i = V_g + A\beta V_i \qquad (8.26)$$

$$\therefore V_i(1 - A\beta) = V_g \qquad (8.27)$$

or,

$$V_i = V_g/(1 - A\beta) \qquad (8.28)$$

But,

$$I_g = V_i/Z_{iA} \qquad (8.29)$$

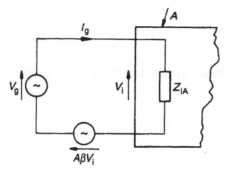

Fig. 8.12 Input impedance with fb is $Z_i (= V_g/I_g)$

Eliminating V_i between Equations (8.28) and (8.29) gives,

$$I_g = V_g/\{Z_{iA}(1 - A\beta)\} \tag{8.30}$$

Let

$$Z_i = V_g/I_g \tag{8.31}$$

Then,

$$Z_i = Z_{iA}(1 - A\beta) \tag{8.32}$$

For A and β given by Equations (8.18) and (8.19), $(1 - A\beta) \simeq (1 + A_o\beta_o)$ for $f \ll f_H$, so Z_i is increased by nfb. Physically, $Z_i > Z_{iA}$ because only a fraction of V_g appears across the amplifier input terminals.

The design of electrometer amplifier circuits exploits the result given by Equation (8.32) and leads to input resistances that are too high ($> 10^{15}\Omega$) to be measured accurately and can only be estimated.

Reduction in output impedance

Figure 8.13 shows an equivalent circuit of the amplifier for the calculation of Z_o, the output impedance with fb. To find Z_o we let $V_g = 0$, apply a generator V_x at the output and find the current I_x that flows. Then,

$$Z_o = (V_x/I_x) \tag{8.33}$$

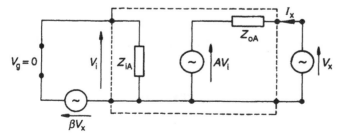

Fig. 8.13 Output impedance with fb is $Z_o (= V_x/I_x)$

Table 8.2 Configurations and parameters of the fb family group

	Feedback type	Configuration	A	β	F	G	Z_i	Z_o	Model
(a)	Series–parallel (shunt)		$A_v = \dfrac{V_o}{V_i}$	$\beta_v = \dfrac{V_{fb}}{V_o}$	$F_v = (1 - A_v\beta_v)$	$G_v =$ Voltage gain $\quad G_v = \dfrac{V_o}{V_g} = \dfrac{A_v}{(1 - A_v\beta_v)}$	$F_v Z_{iA}$	$\dfrac{Z_{oA}}{F_v}$	VCVS
(b)	Series–series		$A_t = \dfrac{I_o}{V_i}$	$\beta_r = \dfrac{V_{fb}}{I_o}$	$F_t = (1 - A_t\beta_r)$	$G_t =$ Transconductance $\quad G_t = \dfrac{I_o}{V_g} = \dfrac{A_t}{(1 - A_t\beta_r)}$	$F_t Z_{iA}$	$F_t Z_{oA}$	VCCS
(c)	Parallel–series		$A_i = \dfrac{I_o}{I_i}$	$\beta_i = \dfrac{I_{fb}}{I_o}$	$F_i = (1 - A_i\beta_i)$	$G_i =$ Current gain $\quad G_i = \dfrac{I_o}{I_g} = \dfrac{A_i}{(1 - A_i\beta_i)}$	$\dfrac{Z_{iA}}{F_i}$	$F_i Z_{oA}$	CCCS
(d)	Parallel–parallel		$A_r = \dfrac{V_o}{I_i}$	$\beta_t = \dfrac{I_{fb}}{V_o}$	$F_r = (1 - A_r\beta_t)$	$G_r =$ Transresistance $\quad G_r = \dfrac{V_o}{I_g} = \dfrac{A_r}{(1 - A_r\beta_t)}$	$\dfrac{Z_{iA}}{F_r}$	$\dfrac{Z_{oA}}{F_r}$	CCVS

Z_{iA}, Z_{oA} are the input and output impedances, respectively, of the forward amplifier without fb.
Z_i, Z_o are the effective impedances with nfb, i.e. $|1 - A\beta| > 1$.

Hence it is called 'series-parallel' fb. Other descriptions are used in the literature: in particular, the word 'shunt' is often employed instead of 'parallel'.

In each configuration the nfb controls the output quantity that is 'sampled' to produce the signal fed back to the input. The means taken to reduce to zero the signal fed back provides a way of classifying fb.

If the fb signal is reduced to zero when the load (Z_L) is (imagined to be) short-circuited then we have 'voltage' fb. Alternatively, if the fb signal is reduced to zero when the load is open-circuited we have 'current' fb. Thus, an alternative description for row (a) is 'series-voltage' fb and the overall performance of the amplifier is summarized by using the circuit-theory description VCVS, i.e. voltage-controlled voltage source.

In general, the dimensions of the overall transfer function, G, give the name to the type of amplifier: e.g. for row (b), $G = G_t = (I_o/V_g)$ so the word 'transconductance' is used for the VCCS (voltage-controlled current source).

For each configuration, G is given by the classic fb formula providing the appropriate meanings are attached to A and β. This can be shown by an analysis similar to that used in the derivation of Equation (8.4a) (*see* Problem 7). In analysis, it is assumed that A and β, have, where appropriate, either zero or infinite input and output impedances. To find the effect of fb on the finite impedances that exist in practice, we can use a procedure similar to that employed in Section 8.3.

The following statements and the word-associations that go with them provide a memory-aid for the effect of nfb on impedance levels, should the reader find it necessary to remember them.

A *series* nfb connection at one of the pairs of amplifier terminals causes an *increased impedance* there, irrespective of the type of nfb connection at the other pair of amplifier terminals. (Word association: the *series* connection of two similar impedances causes an *increased impedance*.)

In contrast, a *parallel* nfb connection at one of the pairs of amplifier terminals causes a *decreased impedance* there, irrespective of the type of nfb connection at the other pair of amplifier terminals. (Word association: the *parallel* connection of two similar impedances causes a *decreased impedance*.)

The factor by which an impedance is changed is $F = (1 - A\beta)$. As an example, consider row (c) of Table 8.2. Without fb, A has an input impedance Z_{iA} and output impedance Z_{oA}. With fb the impedance seen by source, I_g, is $Z_{iA}/(1 - A_i\beta_i)$: the output impedance, Z_o, seen by the load is $Z_{oA}(1 - A_i\beta_i)$.

A circuit implementation, with BJTs, of each of the configurations in Table 8.2 is shown in Table 8.3. Column 2 shows two stages each with a single BJT and 'local' nfb: column 3 shows the application of 'overall' or 'global' nfb, when more than one BJT is used.

To facilitate a comparison of the operation of the circuits with that of their block representations, the necessary input d.c. bias arrangements are omitted from the circuit diagrams.

In each circuit it is possible, in principle, to identify a forward gain function A

Table 8.3

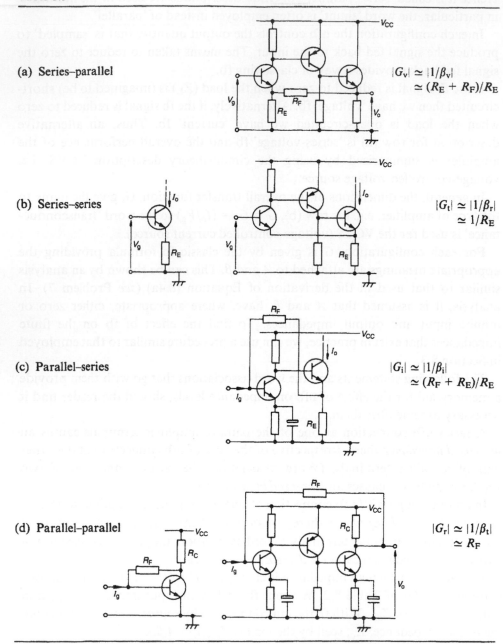

(a) Series–parallel

$|G_v| \simeq |1/\beta_v|$
$\simeq (R_E + R_F)/R_E$

(b) Series–series

$|G_t| \simeq |1/\beta_t|$
$\simeq 1/R_E$

(c) Parallel–series

$|G_i| \simeq |1/\beta_i|$
$\simeq (R_F + R_E)/R_E$

(d) Parallel–parallel

$|G_r| \simeq |1/\beta_t|$
$\simeq R_F$

and a fb fraction β, as is shown in the example that follows. However, in complex circuits involving nfb it is often easier to carry out a straightforward nodal analysis to find the transfer function.

Consider the circuit in row (c) of Table 8.3. An approximate small-signal l.f. equivalent circuit is shown in Fig. 8.15. The nfb loop is broken at the base of the

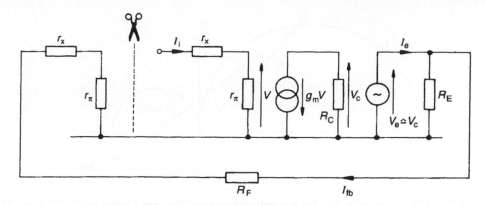

Fig. 8.15   Approximate small-signal equivalent for circuit in Table 8.3, row (c), when nfb loop is cut

input BJT and terminated by the incremental resistance that exists there. At the collector of the input stage, the output stage appears to be an emitter-follower. The simplified representation of this is an ideal voltage amplifier with a gain of +1.

By inspection,

$$V = I_i r_\pi \tag{8.45}$$

$$V_e \simeq V_c = -g_m V R_C = -g_m I_i r_\pi R_C \tag{8.46}$$

For $R_F \gg (r_x + r_\pi)$,

$$I_e = V_e/(R_E \| R_F) \tag{8.47}$$

$$A_i = (I_e/I_i) = -g_m r_\pi R_C/(R_E \| R_F) \tag{8.48}$$

$$\beta_i = (I_{fb}/I_e) = R_E/(R_E + R_F) \tag{8.49}$$

But from Table 8.2,

$$G_i = A_i/(1 - A_i \beta_i) \tag{8.50}$$

Hence, for $-A_i \beta_i \gg 1$

$$G_i \simeq -(1/\beta_i) = -(R_E + R_F)/R_E \tag{8.51}$$

Now, $I_o = \alpha I_e$, $\alpha(\simeq 1)$ being the CB current gain,

$$\therefore (I_o/I_g) = \alpha(R_E + R_F)/R_E \tag{8.52}$$

The approximation $-A_i \beta_i \gg 1$ is valid if, from Equations (8.48) and (8.49),

$$g_m r_\pi (R_C/R_F) \gg 1 \tag{8.53}$$

This condition is easily met since the product $g_m r_\pi$ is the CE current gain of a BJT. Likewise, the other approximations given in the last column of Table 8.3 are normally valid in practice.

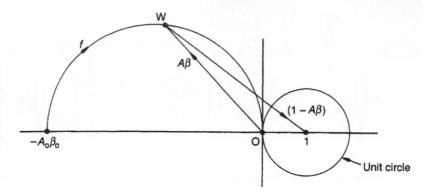

Fig. 8.16 Typical Nyquist plot for a single-stage amplifier: nfb always exists because $A\beta$ lies outside the 'unit circle'

8.5 PROBLEMS WITH NFB

In view of the many benefits offered by the use of nfb it might seem that the greater the amount of fb the better. This is, arguably, true as long as A and β are independent of frequency. However, A, if not β, is *always* to some extent frequency-dependent because of the unavoidable presence of shunt capacitances which cause amplifier phase shift.

Although we can arrange for nfb to exist over an input signal frequency range of interest, the polarity of the fb can change from negative to positive outside that range and that can lead to undesirable effects. A detailed discussion goes beyond the scope of this book. Only a brief qualitative account is given here to introduce the subject. Assume, initially, that A and β have the form given in Equations (8.18) and (8.19). These correspond to a single-stage d.c.-coupled nfb amplifier with a single cut-off frequency. Figure 8.16 shows the semicircular $A\beta$ locus (Nyquist plot). By definition, nfb exists if $|1 - A\beta| > 1$. The contour that describes this is a circle of unit radius centred at the point 1 on the horizontal axis. It is called the 'unit circle'. Outside it nfb exists, inside it there is pfb since $|1 - A\beta| < 1$ (*see* Problem 13). Clearly, the Nyquist plot never enters the unit circle so nfb exists over the whole frequency range.

Suppose now that,

$$A = +A_o/[1 + j(f/f_{H1})][1 + j(f/f_{H2})]$$ (8.54)

and,

$$\beta = -\beta_o$$ (8.55)

These equations are characteristic of a two-stage d.c.-coupled nfb amplifier that has two cut-off frequencies. In this case, the Nyquist plot (Fig. 8.17(a)) is 'cardioid' in shape: it resembles half a heart lying on its side. At N, corresponding to $f = f_N$, the plot enters the unit circle so there is nfb for $f_N > f > 0$ and pfb for $\infty > f > f_N$. Depending on the magnitude of the product $A_o\beta_o$, the pfb can cause

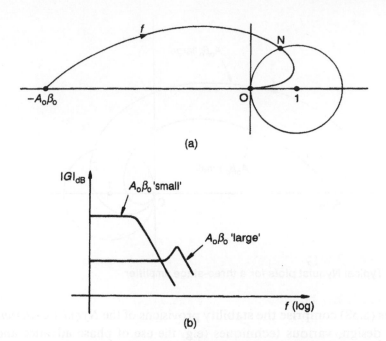

Fig. 8.17 (a) Typical Nyquist plot for a two-stage amplifier. (b) Bode gain plots for (a)

an undesirable peak in the frequency response, as indicated in Fig. 8.17(b), and an associated overshoot followed by a decaying oscillation in the step response of the amplifier. Consider, finally, a nfb amplifier for which,

$$A = -A_o/[1 + j(f/f_{H1})][1 + j(f/f_{H2})][1 + j(f/f_{H3})] \tag{8.56}$$

and,

$$\beta = +\beta_o \tag{8.57}$$

These equations typify a three-stage d.c.-coupled amplifier that has three cut-off frequencies.

The shape of the Nyquist plot (Fig. 8.18) now resembles that of the innermost section of a spiral spring. The plot not only enters the unit circle but also crosses the horizontal axis.

Depending, again, on the product $A_o\beta_o$ it is now possible to pass through, or encircle, the centre point, 1. In that case the amplifier produces a sustained oscillatory output without any externally applied input signal. This condition can be avoided by ensuring that the magnitude of the loop-gain is less than unity when the loop phase shift is zero, i.e.

$$|A\beta| < 1 \tag{8.58a}$$

when,

$$\angle A = 0° \tag{8.58b}$$

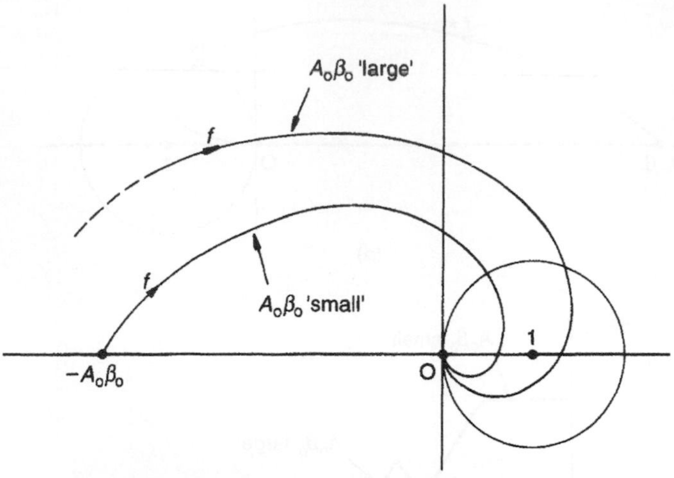

Fig. 8.18   Typical Nyquist plots for a three-stage amplifier

Equations (8.58) comprise the stability provisions of the *Nyquist Criterion*. In nfb amplifier design, various techniques (e.g. the use of phase advance and retard networks) are employed to ensure that the conditions of Equations (8.58) are met and that a satisfactory frequency response is obtained.

8.6   SELF ASSESSMENT TEST

1   Define each of the following terms: forward-gain; closed-loop gain; loop-gain; amount of feedback.

2   How are negative feedback (nfb) and positive feedback (pfb) uniquely defined?

3   By what factor is the gain-tolerance of an amplifier changed when nfb is applied to it?

4   By what factor does the application of nfb increase the bandwidth of a d.c.-coupled amplifier that has a single cut-off frequency?

5   Explain the meanings of the following nfb connection schemes: series–parallel; series–series; parallel–series; parallel–parallel.

6   In what way is the output impedance of an amplifier affected by a series–parallel nfb connection?

7   What limits the maximum amount of nfb that may be applied to a three-stage amplifier?

8.7 PROBLEMS

(Assume throughout, unless otherwise stated, that the amplifier used is a voltage amplifier with series–parallel nfb.)

1   An amplifier with nfb has a gain of $+50$. Without nfb, 50 mV is required to produce a certain output, whereas the input for the same output with nfb must be 1.05 V. Calculate the gain before nfb is added and the feedback fraction.

2   Without nfb, an amplifier has $A = -100 \pm 15\%$. Feedback is applied, with $\beta_0 = 0.04 \pm 2\%$.
    Show that the gain-magnitude, $|G|$, with nfb is, $20.88 \geqslant |G| \geqslant 19.02$.

3   Refer to Fig. 8.7. Show, by geometry, that $MN \equiv |G|$
    (Hint: use similar triangles to show that $|G|/A_0 = [1 - (HN/HJ)]$ and, $(HN/HJ) = |G|\beta_0$)

4   The l.f. gain, $A_{l.f.}$, of an amplifier without fb is,

$$A_{l.f.} = -A_0/[1 - j(f_L/f)]$$

Show that the l.f. gain, $G_{l.f.}$, with nfb and a fb fraction $\beta = +\beta_0$ is

$$G_{l.f.} = -G_0/[1 - j(f'_L/f)]$$

where, $G_0 = A_0/(1 + A_0\beta_0)$ and, $f'_L = f_L/(1 + A_0\beta_0)$

5   An a.c.-coupled amplifier has a gain function A, given by,

$$A = -100/[1 - j(0.04/f)][1 + j(f/500)]$$

where f is expressed in kHz.
Find a similar expression for the gain, G, with nfb if $\beta = +0.03$
(Make sensible engineering approximations.)

6   Without fb an amplifier has an input resistance of 100 kΩ, an output resistance of 100 Ω and a gain function given by,

$$A = -250/[1 + j(10f)]$$

where f is expressed in MHz.
Calculate the new values of input resistance, output resistance and cut-off frequency if nfb is applied and $G = -25$ at very low frequencies.

7   Derive the expressions for G_t, G_i and G_r shown in Table 8.2.

8   In a certain nfb amplifier,

$$A = -100/[1 + \mathrm{j}f]^2$$

where f is expressed in MHz, and $\beta_o = +0.09$.

(a) Show that,

$$|G| = 100/\sqrt{(10 - f^2)^2 + 4f^2}$$

and,

$$\angle G = 180° - \tan^{-1}\{2f/(10 - f^2)\}$$

(b) Show, by differentiation, that $|G|$ has a peak value of 16.7 at $f = 2.83$ MHz.

(c) Draw Bode plots of $|A|$ and $|G|$.

9   For the gain function in 8 show that there is no peaking in the closed-loop response provided $\beta \leqslant 0.01$.

10   A nfb amplifier has,

$$A = +100/(1 + \mathrm{j}f)(1 + \mathrm{j}2f)$$

where f is expressed in MHz, and $\beta = -0.01$.

Calculate $|G|$ at 0.5 MHz, and express the result in polar form.

11   For a nfb amplifier,

$$A = -A_o/[1 + \mathrm{j}(f/f_H)]^3$$

$$\beta = +\beta_o$$

Calculate the value of (f/f_H) at which $\angle A\beta = 0°$ and find the corresponding value of $|A\beta|$.

12   Show that for a nfb amplifier characterized by Equations (8.56), (8.57)

$$\angle A\beta = 0°$$

when $f = \sqrt{(f_{H1}f_{H2} + f_{H2}f_{H3} + f_{H1}f_{H3})}$

(Hint: obtain G in the form $G = -A_o/(a + \mathrm{j}b)$ and equate b to zero.)

13   Show that the curve defined by $|1 - A\beta| = 1$ is a circle of unit radius with its centre at the point 1,0. (Hint: let $A\beta = x + \mathrm{j}y$ and show that $(x - 1)^2 + y^2 = 1$.)

9 Operational amplifiers

During World War II, engineers working on the design of military equipment used thermionic-tube amplifiers in circuits that simulated mathematical operations, e.g. addition, subtraction and integration. Later, the term 'operational amplifier' came to be used in connection with such applications.

Following the introduction of the (Fairchild) microelectronic amplifier type μA709, in 1965, the term 'operational amplifier' or one of its derivative forms, op-amp, OA (used henceforth), has been used to describe not a circuit function but the differential-input, high voltage-gain, direct-coupled IC amplifier itself.

Connected with other components in a feedback loop, the OA has given rise to a very large range of analogue circuits. Costing, now, little more than a few discrete passive components, e.g. resistors, the OA has largely revolutionized bread-and-butter analogue circuit design. It has also enabled engineers of other disciplines (e.g. mechanical, chemical) and scientists, generally, to design circuits having a precision and reproducibility that would previously have required the attention of an experienced electronics specialist.

The aims of this chapter are: to introduce the OA and discuss an ideal model of it for use in first-order circuit analysis and design; to examine some basic circuit applications; and to define some of the most important parameters that characterize deficiencies, of practical OAs, and to consider the ways in which they restrict circuit operation.

9.1 IDEALIZED REPRESENTATION

Figure 9.1(a) is a block schematic representation of an OA and, not being a wiring diagram, it does not show the presence of necessary power supplies, which are dealt with in Section 9.3. The two inputs v_P and v_N and single output v_O are referenced to a common line, i.e. chassis-earth. The letter A_V inside the triangular symbol represents a voltage amplification function and does *not* have a sign associated with it. In this respect, it differs from the 'A' used in the general discussion of feedback in Section 8.1.

The polarity of the gain associated with a signal, in passing through the OA, is indicated by the sign outside the symbol at the input concerned. Thus, v_P is not

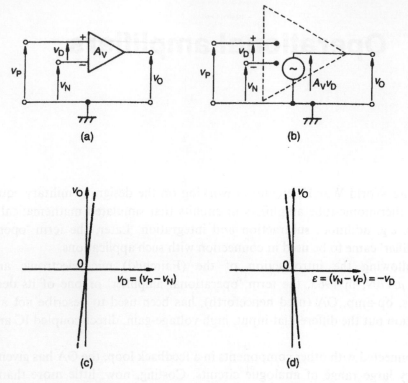

Fig. 9.1 OA representation. (a) Block schematic. (b) Idealized equivalent circuit. (c) Transfer characteristic. (d) Alternative version of (c)

inverted in polarity as it passes through the OA because it is applied at the 'positive' or 'non-inverting' input. In contrast, v_N, applied at the 'negative' or 'inverting' input, is given a change of polarity as it passes through the OA. Note, however, that the '+' and '−' signs are not related to the polarities of the input signals themselves. In Fig. 9.1(a) the '+' sign is shown above the '−' sign. This is an arbitrary drawing choice. When it simplifies circuit recognition, understanding or drawing, the positions of the input terminals may be shown in interchanged positions complete with their signs.

For a first look at the OA we assume it has the simplified circuit model shown inside the dotted triangle in Fig. 9.1(b). The OA is regarded as an idealized voltage controlled voltage source with a frequency-independent gain. The inputs do not load the signal sources in any way, so the input leads are shown open-circuited at the OA input. Furthermore, v_O is independent of any load that may be connected to the output and the direction of the current in it.

The transfer relationship, illustrated in Fig 9.1(c), is,

$$v_O = A_V v_P - A_V v_N = A_V(v_P - v_N) = A_V v_D \tag{9.1a}$$

In this, v_D = differential input voltage.

An alternative, and sometimes more convenient, form of Equation (9.1a) is illustrated in Fig. 9.1(d).

$$v_O = -A_V v_N + A_V v_P = -A_V(v_N - v_P) = -A_V \varepsilon \qquad (9.1b)$$

where, $\varepsilon = (v_N - v_P)$.

It is necessary to emphasize three points in connection with these equations. First, it is the *differential* input signal that is important. The same output is obtained with $v_P = 1\,\text{mV}$ and $v_N = 0\,\text{mV}$ as with $v_P = 1001\,\text{mV}$ and $v_N = 1000\,\text{mV}$. Second, for calculation purposes, the Principle of Superposition applies. The total output is the sum of the component outputs due to each input taken by itself with the other input zero. Third, Equation (9.1) applies to instantaneous signal values, hence the use of lower case symbols with upper case subscripts. DC and sinusoidal signals require a change in notation.

In order to relate OAs to work on amplifiers in previous chapters, sinusoidal quantities will be used in the initial discussion of OA circuit operation and thereafter as appropriate.

Finite-gain analysis

The basic and widely used 'inverter' stage of Fig. 9.2(a) is a convenient starting point. It is an example of an amplifier stage with shunt voltage fb.

The 'stage gain' or 'operational gain' is $G \; (= V_o/V_g)$. Figure 9.2(b) is an equivalent circuit for Fig. 9.2(a). The net loop voltage is $(V_g - V_o)$ and the

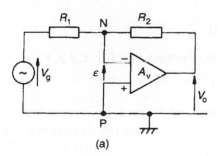

(a)

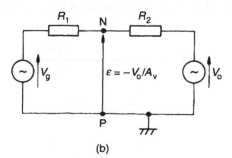

(b)

Fig. 9.2   (a) Inverter configuration with sinusoidal input. (b) Equivalent circuit of (a)

potential difference across R_1 is $(V_g - V_o)R_1/(R_1 + R_2)$. Hence, ε is given by,

$$\varepsilon = V_g - [(V_g - V_o)R_1/(R_1 + R_2)] \tag{9.2}$$

This reduces to,

$$\varepsilon = \alpha V_g + \beta V_o \tag{9.3}$$

where α and β are *not*, in this case, the current gains of a BJT but are given by,

$$\alpha = R_2/(R_1 + R_2) \tag{9.4a}$$

and,

$$\beta = R_1/(R_1 + R_2) \tag{9.4b}$$

Irrespective of the way the amplifier is used as a linear device, it enforces the condition given in Equation (9.1b), i.e.

$$\varepsilon = -V_o/A_v \tag{9.5}$$

Combining Equations (9.3) and (9.5),

$$\alpha V_g + \beta V_o = \varepsilon = -V_o/A_v \tag{9.6}$$

or,

$$\alpha V_g = -\beta V_o[1 + \{1/(A_v\beta)\}] \tag{9.7}$$

$$\therefore G = (V_o/V_g) = -(\alpha/\beta)/[1 + \{1/(A_v\beta)\}] \tag{9.8}$$

or,

$$G = -\alpha A_v/[1 - (-A_v\beta)] \tag{9.9}$$

From Equation (9.8), it is apparent that for $A_v\beta \gg 1$,

$$G \rightarrow -(\alpha/\beta) = -(R_2/R_1) \tag{9.10}$$

Equations (9.6), (9.7) and (9.9) lead to complementary viewpoints on the operating mechanism of the circuit and the significance of Equation (9.10).

Figure 9.3 is phasor representation of Equation (9.6). From this, or Equation (9.7), it follows that if $V_g > 0$ then $V_o < 0$ because $(\alpha, \beta, A_v) > 0$. The condition $A_v\beta \gg 1$ means that the phasor for ε is smaller in magnitude than that for βV_o, so

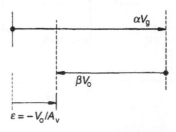

Fig. 9.3   Phasors for Fig. 9.2

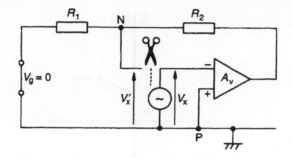

Fig. 9.4   Loop-gain $= (V'_x/V_x)$

αV_g and βV_o almost balance. Equation (9.9) resembles the standard fb equation derived in Chapter 8. However, the coefficient α appears in the expression for gain without fb because of the attenuation suffered by V_g in the input network. The loop-gain is $-A_v\beta$. This can be seen by cutting the loop, as shown in Fig. 9.4: $(V'_x/V_x) = -A_v\beta$, so for $A_v\beta \gg 1$, $G = -(\alpha/\beta) = -(R_2/R_1)$.

Note: the loop-gain defined in Chapter 8 was $A\beta$, but there is no contradiction, because that was *before* the signs of A and β were allocated. In the present case, attaching a sign to 'A' gives $A = -A_v$.

9.2   THE 'VIRTUAL EARTH' CONCEPT

An alternative form for Equation (9.9) is,

$$G = -(\alpha/\beta)[A_v\beta/(1 + A_v\beta)] = -(\alpha/\beta)[1 - \{1/(1 + A_v\beta)\}] \tag{9.11}$$

or,

$$G \simeq -(\alpha/\beta)[1 - (1/A_v\beta)] \tag{9.12}$$

Typically, $A_v > 10^5$ and $\beta > 0.01$ so $A_v\beta > 10^3$. It follows that $|G|$ differs from the ideal value (α/β), i.e. (R_2/R_1), which it would have for the case $A_v\beta = \infty$, by less than one part in 10^3 (i.e. 0.1%). In view of this, it is normal practice in first-order calculations to assume that $A_v\beta = \infty$ or, since β is finite, $A_v = \infty$. This is sometimes known as the 'infinite gain approximation' and leads to the concept of a 'virtual earth'. An OA is considered 'ideal in all respects' if it has the equivalent circuit of Fig. 9.1(b) and a transfer characteristic coincident with the vertical axis (Fig. 9.5). The consequences of this characteristic from a circuit standpoint are shown in Fig. 9.6: $\varepsilon = 0$, whatever the value of V_o.

Point N *appears* to be at the same potential as point P, earth potential in this case, but there is no conducting path between them. N is called a 'virtual earth' point. The concept of a virtual earth is of fundamental importance in the by-inspection analysis of OA circuit operation.

In Fig. 9.6, the existence of a virtual earth at N causes the whole of the input signal V_g to be developed across R_1. R_1 is thus, the overall input resistance of the

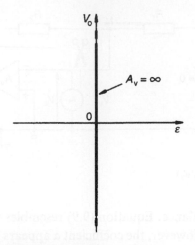

Fig. 9.5   Transfer characteristic for 'ideal' OA

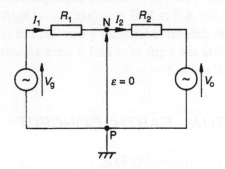

Fig. 9.6   Equivalent circuit of Fig. 9.2(a) with an ideal OA

fb amplifier, and the current in it is V_g/R_1. The potential difference across R_2 is I_2R_2. But, $I_2 = I_1$ and N is at earth potential,

$$\therefore V_o = 0 - I_2R_2 = 0 - I_1R_2 = -V_g(R_2/R_1) \qquad (9.13)$$

or,

$$G = (V_o/V_g) = -(R_2/R_1) \qquad (9.14)$$

An equivalent circuit describing the operation of the fb amplifier is shown in Fig. 9.7, in which $G < 0$.

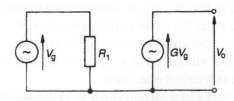

Fig. 9.7   System representation of Fig. 9.6 ($G < 0$)

The virtual earth concept merits further discussion. Referring to Fig. 9.3, it means that $(\alpha V_g + \beta V_o) = 0$, i.e. there is a 'phasor-balancing' mechanism in operation. The sole function of the OA is to produce an output voltage, the magnitude and polarity of which ensure that the balance condition *is* met whatever the nature of the fb component(s).

Figure 9.8 shows a mechanical analogy for the inverter stage under discussion, a rigid beam rigidly pivoted at some point between its ends. If one end moves up a distance h_1 the other end moves down a distance h_2. From the geometry of the diagram, $\sin \theta = (h_1/l_1) = (h_2/l_2)$ so $(h_2/h_1) = (l_2/l_1)$. The analogous quantities are: $h_1 \equiv |V_g|$; $h_2 \equiv |V_o|$; $l_1 \equiv R_1$; $l_2 \equiv R_2$.

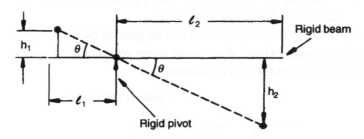

Fig. 9.8   'See-saw' analogy of an OA inverter

OA inverter stages using thermionic-tube amplifiers were sometimes called 'See-saw' amplifiers.

If R_1 is replaced by an impedance Z_1 and R_2 by an impedance Z_2 then, by reasoning identical to that leading to Equation (9.14), it follows that the general inverter gain formula is,

$$G = (V_o/V_g) = -(Z_2/Z_1) \tag{9.15}$$

Example 9.1

Design an inverter stage for which $G = -5$ and the input resistance is $10\,\text{k}\Omega$. Comment on the effect of resistor tolerances on gain accuracy.

Solution

Referring to Fig. 9.7, we require $R_1 = 10\,\text{k}\Omega$. From Equation (9.14), $G = -(R_2/R_1)$: hence, for $G = -5$, $R_2 = 50\,\text{k}\Omega$. The choice of two $100\,\text{k}\Omega$ resistors in parallel avoids the use of special value resistors.

Suppose R_1 and R_2 both have fractional tolerances of magnitude δ ($\ll 1$). Then, $R_1 = R \pm R\delta = R(1 \pm \delta)$: similarly, $R_2 = 5R(1 \pm \delta)$. In one worst-case condition, R_2 is high and R_1 low.

$$\therefore |G(\text{max})| = 5(1 + \delta)/(1 - \delta) = 5[1 + \{2\delta/(1 - \delta)\}] \simeq 5(1 + 2\delta)$$

We can arrive at Equation (9.20) by a different route, Superposition. Thus

$$v_O = (v_{O1} + v_{O2} \cdots + v_{On}) \tag{9.21}$$

where v_{Or} is the component part of v_O due to v_r acting alone, i.e. with all the other inputs set at zero. In this condition i_1 is due to v_r because the potential difference across each of the other resistors is zero. The circuit acts as a simple inverter for the rth input so using Equation (9.14) (or Equation (9.15) if impedances, generally, are used),

$$v_{Or} = -(R_F/R_r)v_r \tag{9.22}$$

Substituting for v_{O1}, v_{O2}, etc., in Equation (9.21) gives Equation (9.19).

The inverting summer is a useful circuit for the (linear) mixing of analogue waveforms. Also the design of 'weighted resistor' digital-analogue converters is based on Fig. 9.10(a): in that case, $R_2 = 2R_1$, $R_3 = 2R_2$, etc.

The input circuit for the rth bit is shown in Fig. 9.10(b). Logic controlled bit-switch S_r gives $v_r = V_R$ ($\equiv$'1') and a corresponding input current contribution V_R/R_r in position 'a' and $v_r = 0$ ($\equiv$'0') and zero contribution to i_1 in position 'b'.

(b) The integrator
Figure 9.11(a) shows an integrator. S symbolizes the action of an electronic switch. When S is closed, $v_O = \varepsilon = 0$ irrespective of v_G. When S opens (at $t = 0$, say),

$$i_1 = v_G/R \tag{9.23}$$

and,

$$i_2 = C[d(0 - v_O)/dt] = -C(dv_O/dt) \tag{9.24}$$

Equating i_1 and i_2,

$$v_O = -(1/CR) \int_0^{t_d} v_G dt \tag{9.25}$$

In a popular application, the generation of linear timebase sweeps (for oscilloscopes and precision analogue timer circuits), $v_G = -V$, a d.c. level derived from a zener diode operating in the breakdown region. Then, when S opens, Equations (9.23) and (9.24) give,

$$(dv_O/dt) = +(V/CR) = \text{constant.} \tag{9.26}$$

It is arranged that S closes after a specified time interval or after a prescribed level of v_O is reached. The resultant 'run-up' waveform is shown in Fig. 9.11(b).

(c) The non-inverting amplifier
Alternative ways of representing the non-inverting amplifier are shown in Fig. 9.12(a) and (b).

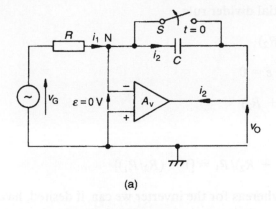

(a)

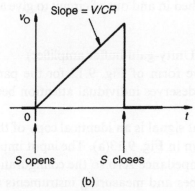

(b)

Fig. 9.11   (a) An OA integrator. (b) Timebase sweep for the case $v_G = -V$

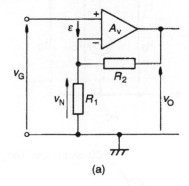

(a)

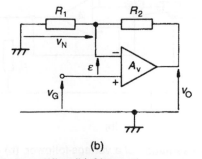

(b)

Fig. 9.12   (a) A non-inverting amplifier. (b) Alternative representation of (a)

From the potential divider rule,

$$v_N = v_O R_1/(R_1 + R_2) \qquad (9.27)$$

But, $v_N = v_G$ since $\varepsilon = 0$

$$\therefore v_G = v_O R_1/(R_1 + R_2) \qquad (9.28)$$

or,

$$G = (v_O/v_G) = (R_1 + R_2)/R_1 = [1 + (R_2/R_1)] \qquad (9.29)$$

Note that $G \geqslant 1$, whereas for the inverter we can, if desired, have $|G| < 1$. Values of R_1, or R_2, can be switched in and out of circuit to give a switched-gain amplifier (*see* Problem 7).

(d) The voltage-follower (Unity-gain buffer amplifier)
This is really a derivative form of Fig. 9.12 for the particular cases $R_2 = 0$ or $R_1 = \infty$, or both, but it deserves individual attention because it is a very widely used analogue block.

$G = +1$, i.e. the output signal is an identical copy of the input signal. An ideal equivalent circuit is shown in Fig. 9.13(a). The input impedance is (theoretically) infinite and the output impedance zero so the configuration is useful in the input stages of circuit test probes and measuring instruments (*see* Problem 8). Figure

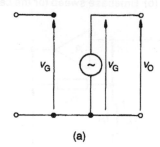

(a)

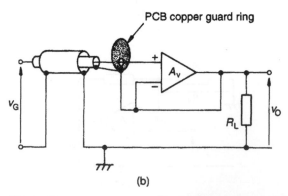

(b)

Fig. 9.13   (a) Simplified representation of a voltage-follower. (b) A voltage-follower and its application in electrometry

9.13(b) shows the signal fed in via a double-screened cable, such as might be used in electrometry.

The output of the OA drives a printed circuit board (PCB) copper guard ring surrounding the input lead and also drives the inner screen of the cable. The principle employed here is that there can be no leakage current between two points when their potentials are always equal. The earthed outer screen of the cable minimizes electrostatic pick-up effects: any leakage current between the screens is supplied by the OA and not by the input signal.

(e) The difference amplifier (or 'differential' amplifier)
This is shown in Fig. 9.14(a) in a typical application, as a thermocouple amplifier giving an output, v_O, that is proportional to the difference between the thermo-electric voltages, v_A and v_B, produced in the thermocouple elements.

Figures 9.14(b) and (c) show how the circuit can be analysed using Super-position. With, $v_B = 0$, the circuit behaves as an inverter so the output voltage component, v_{O1}, due to v_A is given by,

$$v_{O1} = -av_A \tag{9.30}$$

With $v_A = 0$, the signal appearing at the non-inverting input is $av_B/(1 + a)$. The circuit now behaves as a non-inverting amplifier with gain $[(R + aR)/R] = (1 + a)$.

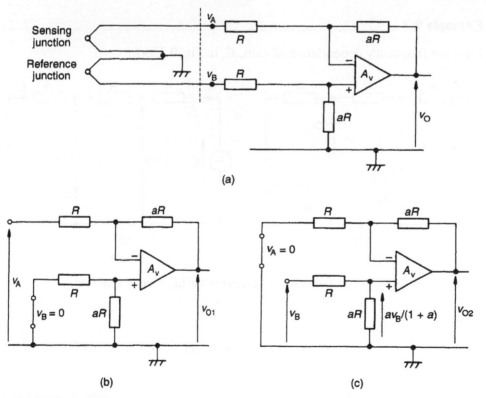

(a)

(b)   (c)

Fig. 9.14   (a) A 'difference' amplifier used with a thermocouple: (b), (c) show circuit conditions for applying Superposition

Hence,

$$v_{O2} = +av_B(1 + a)/(1 + a) = av_B \tag{9.31}$$

and,

$$v_O = v_{O1} + v_{O2} = a(v_B - v_A) = -a(v_A - v_B) \tag{9.32}$$

The circuit may be regarded as a subtractor.

Example 9.2

Design a non-inverting stage having $G = +4 \pm 2\%$ and an input resistance of $1\ M\Omega \pm 10\%$.

Solution

From Equation (9.29), $G = (R_1 + R_2)/R_1$. For $G = 4$, $R_2 = 3R_1$, i.e. $R_1 = R$, $R_2 = 3R$. The choice of R is to some extent arbitrary (but see Section 9.3). Let $R_1 = 10\ k\Omega$, then $R_2 = 30\ k\Omega\ (= 18\ k\Omega + 12\ k\Omega)$.

The tolerance requirement is met using $\pm 1\%$ resistors. To obtain the required input resistance, connect a $1\ M\Omega \pm 10\%$ resistor from the non-inverting terminal to the earth.

Example 9.3

Find the frequency dependence of gain, G, in Fig. 9.15(a).

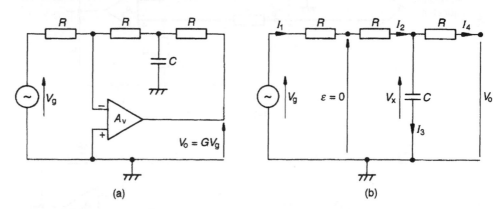

(a)   (b)

Fig. 9.15   (a) Circuit for Example 9.3. (b) Equivalent of (a), used in analysis

Solution

Refer to Fig. 9.15(b), in which V_x is the potential difference across C.
By inspection, $I_1 = V_g/R = I_2$

$$V_x = 0 - I_2R = -(V_g/R)R = -V_g$$

$$I_3 = V_x/(1/j\omega C) = -j\omega C V_g$$

$I_4 = I_2 - I_3 = (V_g/R) + j\omega CV_g$

$V_o = V_x - I_4R$

$\therefore V_o = -V_g - [(V_g/R) + j\omega CV_g]R$

$\therefore G = (V_o/V_g) = -2[1 + j(\omega CR/2)]$

Note: in this problem the fb network is a three-terminal circuit in place of a single two-terminal component, but that does not alter the method of attack.

Non-linear configurations

Employing components with non-linear *I–V* characteristics in the fb loop leads to some very useful analogue circuit designs, three of which are dealt with next. The procedure given earlier is still applicable.

(a) Logarithmic amplifier (Fig. 9.16(a))
The OA enforces the condition $V_{CB} = 0$, so $I_C = V/R$.
But, from Chapter 3 (Equation (3.1)), for $V_{BE} > 100$ mV,

$$I_C = I_S e^{V_{BE}/V_T} \tag{9.33}$$

Hence, the OA output voltage assumes a value that satisfies Equation (9.33). Thus

$$V_O = -V_{BE} = -V_T \log_e(I_C/I_S) \tag{9.34}$$

The load line construction for V_{BE} is shown in Fig. 9.16(b). The logarithmic amplifier is the basis of precision analogue IC multiplier and exponential function generator designs.

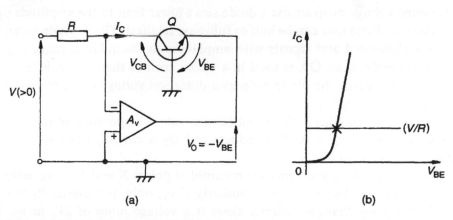

Fig. 9.16 (a) A basic logarithmic amplifier. (b) Load line construction for (a)

(b) Precision current generator (Fig. 9.17)
By inspection, $V_N = V_P = V_Z$: hence, $I_E = V_Z/R_E$. Allowing for the CB current gain ($\geqslant 0.99$) of Q,

$$(V_Z/R_E) > I_C \geqslant 0.99(V_Z/R_E) \tag{9.35}$$

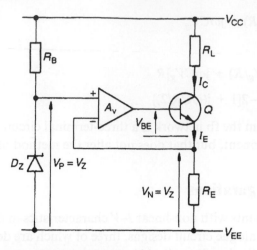

Fig. 9.17   A precision d.c. current generator

V_{BE} takes on that value required to sustain this, so the output voltage of the OA rises to $(V_{EE} + V_Z + V_{BE})$. Note that I_C is sensibly independent of BJT parameters, V_{EE} and V_{CC}, provided $(V_{CC} - I_C R_L) > (V_{EE} + V_Z + V_{BE})$. Furthermore, if D_Z is appropriately chosen, I_C is almost independent of temperature, T.

Tolerances on V_Z and R_E can be trimmed out by connecting a potentiometer in series with R_E.

Voltage-frequency converters and current-output digital-analogue converters employ circuits similar to that in Fig. 9.17.

(c) Precision half-wave rectifier

The forward voltage drop across a diode sets a lower limit to the amplitude of a sinusoidal waveform that can be half or full-wave rectified. The diode drop can be effectively eliminated and signals with amplitudes in the millivolt range can be rectified, precisely, if an OA is used in a circuit such as that of Fig. 9.18(a). In effect, the OA regards the diode drop as a distortion voltage in the fb loop and deals with it accordingly.

The waveforms of Figs 9.18(b) and (c) illustrate the operation of the circuit. Summarizing: when $v_G > 0$, D_1 conducts and D_2 is cut off; when $v_G < 0$, D_2 conducts and D_1 is cut off.

Half-wave rectified waveforms are obtained at points X and Y. If v_X, only, is required R_3 can be dispensed with: similarly, if v_Y, only, is required R_2 can be omitted. When v_G changes polarity there is a voltage jump of $2V_\gamma$ in v_O, V_γ ($\simeq 0.6$ V) being the threshold-of-conduction voltage of D_1, D_2.

In a full-wave rectifier development of Fig. 9.18(a) the waveforms v_X, v_Y are combined in a differential amplifier stage (not shown).

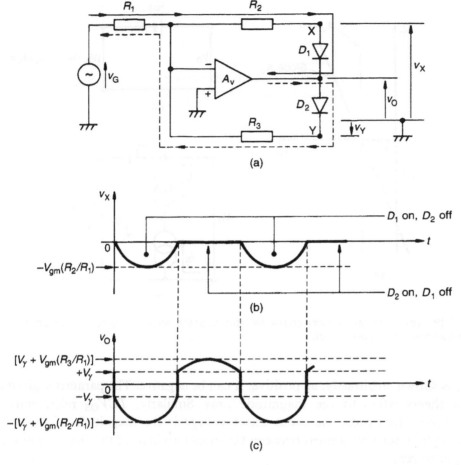

Fig. 9.18   (a) Precision half-wave rectifier. (b) Waveform at X (full line shows current for $v_G > 0$, dotted line for $v_G < 0$). (c) Waveform at OA output

9.3   THE NON-IDEAL OA

Some of the more important non-ideal parameters of practical OAs are defined in this section and the effects they have on some of the standard linear configurations, already discussed, are analysed.

Static defects

Figure 9.19(a) is an exaggerated view of the transfer characteristic of a practical OA in the vicinity of the origin. In general $V_O \neq 0$ when $V_D = 0$. There is a small 'input offset voltage', V_{OS} (or V_{IO}): it is standard and convenient to specify this rather than an 'output offset voltage'.

The equation for the transfer characteristic is,

$$V_O = A_V(V_D - V_{OS}) \tag{9.36a}$$

A normal design choice is $R_3 = R_{eq}$. Then $(I_N R_{eq} - I_P R_{eq}) = \pm I_{OS} R_{eq}$: that is the reason for including R_3 in an inverter stage in which $V_{G2} = 0$.

Then, substituting for α and β, Equation (9.39) gives,

$$V_O = [V_{G2}(R_1 + R_2)/R_1] - [V_{G1}(R_2/R_1)] \pm [(I_{OS} R_{eq} + V_{OS})(R_1 + R_2)/R_1] \quad (9.40)$$

The first two terms on the right hand side of Equation (9.40) give the ideal output. The last term gives the worst case error due to the combined effects of V_{OS} and I_{OS}. A '$\pm$' sign choice is now used to allow for the fact that the polarities of these parameters are not known.

Note that the total offset error is the same whether the circuit is used as an inverter, or as a non-inverter, or as a difference amplifier. To calculate integrator offset errors refer to Fig. 9.21.

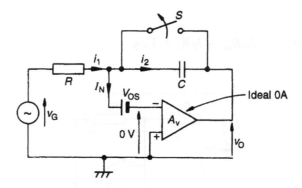

Fig. 9.21   DC defects in an integrator

When S is closed, $v_O = V_{OS}$ so $i_1 = (v_G - V_{OS})/R$. When S opens,

$$i_2 = (i_1 - I_N) = C[\mathrm{d}(V_{OS} - v_O)/\mathrm{d}t] \quad (9.41)$$

Making the usual first-order approximation $V_{OS} \neq f(t)$, Equation (9.41) gives,

$$[(v_G - V_{OS})/R] - I_N = -C(\mathrm{d}v_O/\mathrm{d}t) \quad (9.42)$$

$$\therefore v_O = \pm V_{OS} - (1/CR) \int [v_G - I_N R \pm V_{OS}]\, \mathrm{d}t \quad (9.43)$$

The first term on the right hand side of Equation (9.43) represents the initial output condition. Again, for worst case error the '$\pm$' sign option is used for the offset term.

Dynamic defects

These relate to the small-signal frequency response and large-signal step and sinusoidal responses of the OA. Figure 9.22(a) shows a small-signal equivalent circuit: Z_{cm} = common-mode input impedance; Z_{dm} = differential-mode input impedance; Z_{oA} = output impedance; $A_v(jf)$ = frequency-dependent voltage gain.

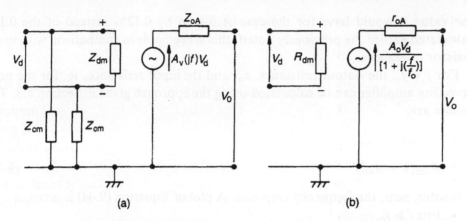

Fig. 9.22   (a) Small-signal equivalent circuit of practical OA. (b) Simplified form of (a)

In general, the impedances are complex. However, for an assessment of their effects in the audio frequency range they may be approximated adequately by their resistive parts. In the simplified equivalent circuit of Fig. 9.22(b), R_{cm} ($\equiv Z_{cm}$) is ignored because of its very high value ($>10\,\text{M}\Omega$) in comparison with other circuit component values.

$R_{dm}(\equiv Z_{dm}) \sim 1\,\text{M}\Omega$ for bipolar OAs, but may be ignored with FET OAs because it is so large ($>100\,\text{M}\Omega$). Typically, $r_{oA} \sim 100\,\Omega$. To minimize the possibility of closed-loop stability problems the OA is often designed so that, over the usable frequency range,

$$A_v(jf) = A_o/[1 + j(f/f_o)] \tag{9.44}$$

where, A_o is the gain for $f \ll f_o$.

We now consider the effect of R_{dm}, r_{oA} and calculate the frequency response of the inverter and non-inverter, using Equation (9.44). Figure 9.23 shows an equivalent circuit for the calculation of loop-gain when R_{dm}, r_{oA} are accounted for. R_{dm} is in parallel with R_1 and r_{oA} in series with R_2. Hence, if we make $R_1 < (R_{dm}/10)$ and $R_2 > 10r_{oA}$, then the loop-gain differs from that with $R_{dm} = \infty$ and $r_{oA} = 0$ by less than 20%. Using the figures assumed at the start of Section 9.1, this means that the closed-loop gain-magnitude, $|G|$, differs from

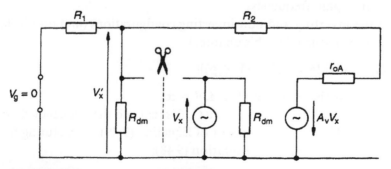

Fig. 9.23   Equivalent circuit for loop-gain calculation

the value it would have for the case $A_v\beta = \infty$ by 0.12% instead of the 0.1% calculated earlier. As previously noted, this is negligible in comparison with usual resistor tolerances.

For $f \ll f_o$, the output resistance, r_o, and the input resistance, R_i, for the non-inverting amplifier can be calculated using the approach given in Section 8.3. The results are,

$$R_o = r_{oA}/(1 + A_v\beta) \qquad (9.45)$$

$$R_i = R_{dm}(1 + A_v\beta) \qquad (9.46)$$

Consider, next, the frequency response. A plot of Equation (9.44) is given in Fig. 9.24. For $f \gg f_o$,

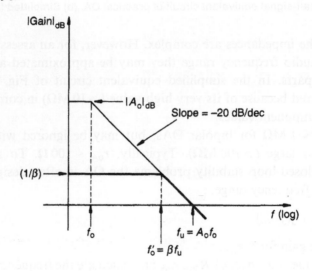

Fig. 9.24   OA frequency response on open and closed-loop

$$|A_v(jf)| = A_o/(f/f_o) \qquad (9.47)$$

or,

$$|A_v(jf)|f = A_of_o = f_u \qquad (9.48)$$

where f_u = unity-gain frequency.

For both the inverting and non-inverting configurations (and for all the linear OA applications discussed in this chapter),

$$|G(\text{practical})| = |G(\text{ideal})|/|[1 + (1/A_v\beta)]| \qquad (9.49)$$

where, $|G(\text{ideal})|$ is the stage gain for $A_v\beta = \infty$.

When resistors are used to determine G, β is real and the frequency dependence of G is determined by the denominator of Equation (9.49). Substituting for $A_v(jf)$ from Equation (9.44) and using Equation (9.48),

$$(1/A_v\beta) = [1 + j(f/f_o)]/A_o\beta = (1/A_o\beta) + j(f/f_u\beta) \qquad (9.50)$$

Since $1 \gg (1/A_o\beta)$, it follows that,

$$[1 + (1/A_v\beta)] \simeq [1 + j(f/f_u\beta)] \tag{9.51}$$

The closed-loop cut-off frequency occurs at,

$$f = f_o' = f_u\beta \tag{9.52}$$

For the non-inverting amplifier, $|G_{l.f.}| = 1/\beta$. Hence

$$|G_{l.f.}| \times f_o' = f_u \tag{9.53}$$

Thus, the gain-bandwidth product is constant.

For the inverting amplifier, $|G_{l.f.}| = (R_2/R_1) = [(1/\beta) - 1]$

$$\therefore [|G_{l.f.}| + 1]f_o' = f_u \tag{9.54}$$

Note that the bandwidth of a unity-gain inverter is one half that of a unity-gain non-inverter because $\beta = (1/2)$ in the former case and $\beta = 1$ in the latter.

To conclude this introduction to OA defects we consider large-signal behaviour. 'Slew rate' is defined by reference to the voltage-follower test circuit of Fig. 9.25(a). For a large (e.g., 10 V) step change in input voltage the input stage of an OA, of the type considered in this chapter, cuts off and the amplifier operates in a non-linear manner. (The small-signal equivalent circuit is no longer applicable.)

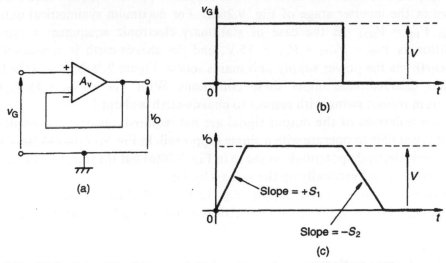

Fig. 9.25 (a) Test circuit for 'slew rate'. (b) Input waveform. (c) Output waveform

The slew rate, S, is the maximum rate of change of v_O under these conditions. Referring to Fig. 9.25(b) and (c) it is either S_1 or S_2, whichever is the smaller, and is expressed in V/μ s.

The slew rate sets a limit to output voltage amplitude at a given frequency. For sinusoidal signals,

$$v_o = V_{om} \sin \omega t \tag{9.55}$$

$$\therefore (dv_o/dt) = \omega V_{om} \cos \omega t = 2\pi f V_{om} \cos \omega t \tag{9.56}$$

The maximum rate of change of v_o occurs when cos $\omega t = 1$.

$$\therefore S = 2\pi f V_{om} \tag{9.57}$$

or,

$$(V_{om} \times f) = S/2\pi \tag{9.58}$$

The 'full-power bandwidth' is that f for which V_{om} is the maximum possible with the power supplies used.

9.4   PRACTICAL POINTS

This section considers, briefly: power supply connection; d.c. feedback provision; types of OA; and component choice.

Supply rail connections

For the output voltage, v_O, to be able to take on values both positive and negative with respect to chassis-earth, two power supplies V_{PP} and V_{NN} are required, as shown in the inverter stage of Fig. 9.26(a). For maximum symmetrical output swing $V_{PP} = V_{NN}$. In the case of stationary electronic equipment a typical condition is $V_{PP} = V_{NN} = V_{CC} = 15$ V, and the chassis-earth is connected to true-earth via the power supply unit mains socket. Figure 9.26(b) shows the OA transfer characteristic under these conditions. With $V_{PP} = V_{NN} = 15$ V, the maximum output swing with respect to chassis-earth is about 13.5 V.

If two polarities of the output signal are not required, under d.c. conditions, then it is possible to operate with a single supply rail. In Fig. 9.27 the OA 'sees' the same inter-electrode potentials as those in Fig. 9.26(a) but the transfer characteristic is now shifted vertically up the v_O axis by V_{NN}.

To operate from a single supply, e.g. 15 V we let $(V_{PP} + V_{NN}) = 15$ V and arrange that $V_{NN} = 7.5$ V for a symmetrical output swing (*see* Problem 17).

Feedback precautions

It is important to remember that whatever the intended function of an OA, connected to work as a linear amplifier, there *must always* be a d.c. path between the OA output and the inverting input terminal in order to establish stable d.c. operating conditions. Normally this path is present all the time but in some applications it may be switched in and out, as is the case with an integrator resetting switch. It follows that the circuit of Fig. 9.28 is completely unacceptable, as it stands. It could work if a (large) resistor were connected in parallel with the fb combination.

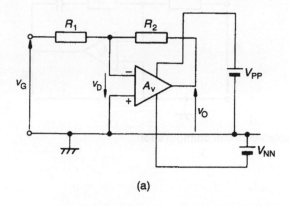

(a)

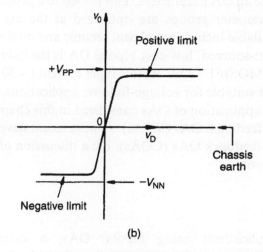

(b)

Fig. 9.26 (a) Inverter with dual power supplies. (b) Transfer characteristic (not to scale)

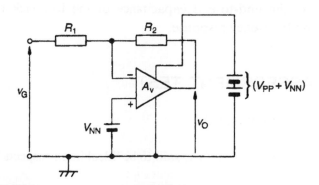

Fig. 9.27 Rearrangement of Fig. 9.26(a) for single supply operation

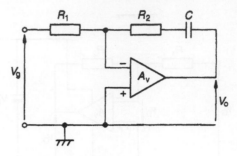

Fig. 9.28   Unacceptable circuit configuration

OA Types

Because of the nature of parameter inter-relationships it is not possible simultaneously to optimize all OA parameters. This has led to a profusion of OA types, in which specific parameter groups are optimized at the expense of the others. Package types available include metal can, ceramic and dual-in-line.

A popular, multi-sourced, low-cost bipolar OA is the industry-standard '741'. The CA3140 is a (MOS)FET OA. Its low input current (~ 50 pA) and high input impedance make it suitable for voltage-follower applications.

The theory and application of OAs considered in this chapter relates to a class known as voltage-feedback OAs (VOAs). A more recent development is the class known as current-feedback OAs (COAs), but a discussion of those goes beyond the scope of this book.

Resistor choice

For inverter applications using bipolar OAs, a desirable condition is $R_{dm} \gg R_1 \gg R_g$. This requirement is usually met with $R_{dm} \gg R_1 \geqslant 10$ kΩ. A consequent limitation on R_2 is usually, 1 M$\Omega \geqslant R_2$ otherwise the closed-loop bandwidth is confined to the low audio range.

With FET OAs an upper limit to R_2 is set by closed-loop bandwidth requirements. With the very large values of R_2 (e.g. 10^{10} Ω) used in some current-voltage converters the end-to-end capacitance of the feedback resistor has a dominant effect on frequency response.

9.5   SELF ASSESSMENT TEST

1 What is the definition of an OA?

2 What are the properties of an OA, said to be 'ideal in all respects'?

3 What is meant by 'virtual earth'?

4 Which method of circuit analysis lends itself better to OA configurations, mesh or nodal?

5 As an OA voltage-follower gives a voltage gain of unity why not replace it by a piece of wire joining the input and output terminals?

6 What is meant by the following terms: input offset voltage; input offset current; unity-gain frequency; slew rate?

7 Why is it not acceptable to use an OA without nfb in amplifier applications?

8 Why use two supply rails for an OA when it can operate with one?

9.6 PROBLEMS

In Problems 1 to 17 assume that the OAs used are ideal in all respects. Remember that in design problems there are rarely, if ever, *unique* solutions. The component values given for them in the 'Answers' section are indicative only.

1 Design an OA inverter having $G = -10 \pm 2\%$, $R_i = 10 \text{ k}\Omega \pm 5\%$.

2 Derive an expression for v_O for the adder-subtractor circuit of Fig. 9.29. Hence, choose resistor values to give,

$$v_O = v_1 + 5v_2 - 2v_3 - 3v_4$$

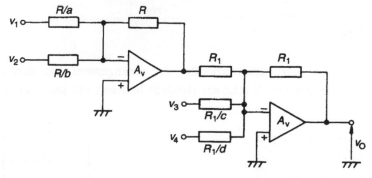

Figure 9.29

3 The 'T-type' fb network of Fig. 9.30 permits high values of voltage gain to be obtained without the use of very large resistors. Show that,

$$G = (-R_2/R_1)[1 + (R_4/R_3)]$$

Hence, design for $G = -100 \pm 2\%$, $R_i = 100 \text{ k}\Omega \pm 5\%$.

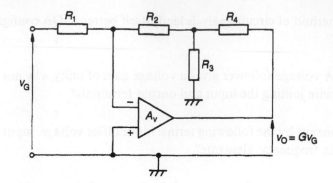

Figure 9.30

4   Show that for the meter-multiplier circuit of Fig. 9.31,

$$I_M = I[1 + (R/r)]$$

Meter M requires $I_M = 50 \, \mu A$ for full scale deflection (FSD). Design for FSD when $I = 1 \, \mu A$.

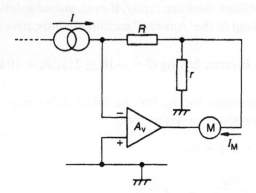

Figure 9.31

5   Figure 9.32 shows, schematically, an arrangement for determining the magnetic flux density B (shown shaded) between the poles of a permanent

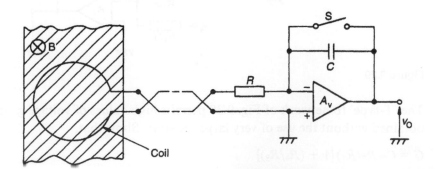

Figure 9.32

magnet. A small coil of n turns and mean area A_S is positioned perpendicular to the direction of the flux. At $t = 0$, switch S is opened and the coil is removed from the vicinity of the magnet. Show that,

$$B = CR|\Delta v|/nA_S$$

where $|\Delta v|$ is the magnitude of the voltage change at the output of the OA. (Hint: |induced emf| = |rate of change of flux linkages|)

6 Design a non-inverting amplifier for $G = +5 \pm 2\%$, $R_i = 500\ k\Omega \pm 5\%$.

7 Design the switched-gain amplifier of Fig. 9.33 for: $G = +10 \pm 2\%$, S closed; $G = +2 \pm 2\%$, S open.

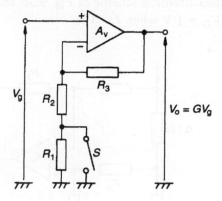

Figure 9.33

8 In Fig. 9.34, the reading of voltmeter V changes by 1 V when S is switched from 'a' to 'b'. Calculate the resistance of the meter on the (10 V) range used.

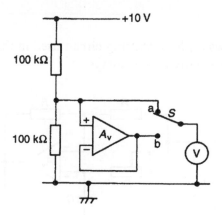

Figure 9.34

9 Explain the operation of the polarity-switched amplifier of Fig. 9.35.

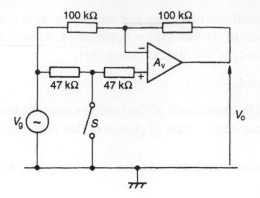

Figure 9.35

10 In the current-measurement scheme of Fig. 9.36, choose values of R_1, R_2, R_3 and R_4 so that $V_O = 1$ V when $I = 1$ A.

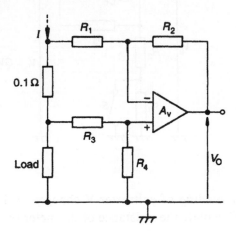

Figure 9.36

11 Figure 9.37 shows a phase-shifting circuit used in the generation of circular symbols for cathode ray tube displays.

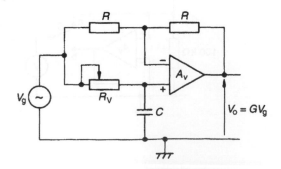

Figure 9.37

Show that, for sinusoidal input signals,

$$G = (1 - j\omega CR_X)/(1 + j\omega CR_X)$$

where R_X is that part of R_V not shorted-out.
Calculate R_X for a phase shift of 90° if $R = 10\,k\Omega$, $C = 10\,nF$ and V_g is a 1 kHz signal.

12 Calculate I_O for the low-current generator of Fig. 9.38.

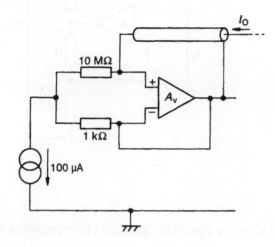

Figure 9.38

13 Figure 9.39 shows an active bridge using four resistance strain gauges, R_1, R_2, R_3 and R_4.

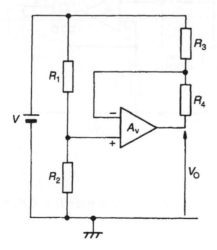

Figure 9.39

Calculate V_O, if $R_1 = R_2 = R_3 = R$ and $R_4 = R(1 + \alpha)$

14   Calculate v_O for the circuit of Fig. 9.40.

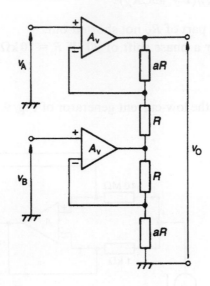

Figure 9.40

15   Figure 9.41 shows a circuit (a 'gyrator') for simulating an inductance.

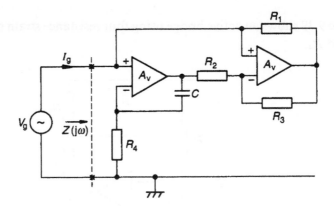

Figure 9.41

Show that $Z(j\omega) = (V_g/I_g) = j\omega L$ where $L = CR_1R_2R_4/R_3$

16   In the instrument rectifier scheme of Fig. 9.42, M is a moving coil d.c. meter. Calculate R, if M is to indicate an FSD of 50 μA when v_G is a 1 kHz sinusoidal waveform, balanced about earth, with an RMS value of 1 V.

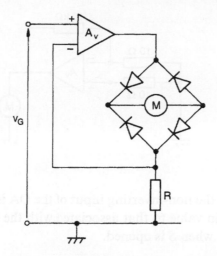

Figure 9.42

17 The audio amplifier in Fig. 9.43 operates from a single supply rail.

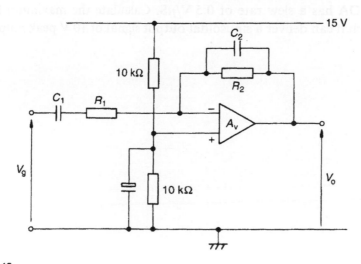

Figure 9.43

Show that,

$$G = (V_o/V_g) = -G_o/[1 - j(f_L/f)][1 + j(f/f_H)]$$

where $G_o = (R_2/R_1)$; $f_L = 1/2\pi C_1 R_1$; $f_H = 1/2\pi C_2 R_2$.
Design a stage giving: $G_o = 10 \pm 2\%$; $f_L = 20$ Hz; $f_H = 20$ kHz; mid-band
input resistance > 10 kΩ.

18 Figure 9.44 shows a test circuit for finding the offset voltage of an OA using a
low cost moving coil voltmeter, M. Calculate V_{OS} if $V_O = 1.22$ V.
(Make sensible approximations and state what they are.)

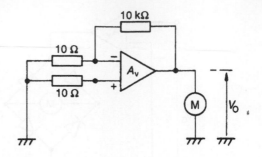

Figure 9.44

19 If, in Fig. 9.21, the non-inverting input of the OA is connected to earth via a resistor, equal in value to that associated with the inverting input, calculate the effect on v_O when S is opened.

20 An inverter stage uses an OA having $f_u = 1\,\text{MHz}$. Calculate the gain, in polar form, at 200 kHz if the gain at very low frequencies is -10.

21 An OA has a slew rate of 0.5 V/μS. Calculate the maximum frequency at which it can deliver a sinusoidal output signal of 10 V peak amplitude.

10 The BJT and FET as switches

The BJT and FET have been regarded so far, in this book, primarily as small-signal amplifying devices. However, it is necessary for the analogue electronics engineer to understand the operation and limitations of these devices as *switches* in order to be able to select or, if need be, design circuits that form the 'interface' between the analogue and digital sub-systems that normally comprise a complete electronic system.

The BJT and FET are each capable of functioning as a 'voltage switch' or as a 'current switch'. With the voltage switch, an applied or circuit-defined voltage is switched in, or out, of circuit according to the state of a control signal applied to the device input terminals.

With the current switch, the control signal switches an existing, circuit-defined, current through the device or through an alternative path in the circuit. The current switch is superior to the voltage switch with respect to attainable switching speed but it wastes power and is more complicated to use.

The voltage switch dissipates a relatively small power, is straightforward to design and use, and is more likely to be encountered by a newcomer to the field. Figure 10.1(a) illustrates the function of a voltage switch. v_S is a voltage source and may be time-dependent or constant: the series-connected ON–OFF switch S is actuated by a two-state signal supplied from a control circuit.

The ideal characteristics of S are shown in Fig. 10.1(b). Thus, whatever the magnitude and polarity of v_S, the potential difference, v, across the terminals of S

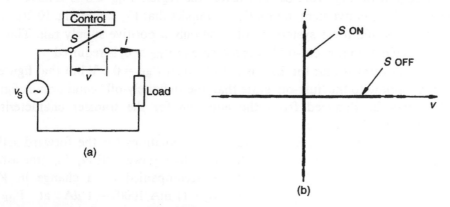

(a)

(b)

Fig. 10.1 (a) A series connected voltage switch. (b) Ideal ON and OFF characteristics

is zero when the contacts are closed and the current, i, is zero when the contacts are open. Additional properties of an ideal switch are: the switching time from OFF to ON, or vice-versa, is zero; the control signal dissipates no power. The extent to which a BJT and an FET possess these ideal characteristics is discussed in this chapter.

For each type of device, simple d.c. models are presented for the OFF and ON conditions. This is followed by a discussion of dynamic performance and of some typical applications. In conclusion, a section is devoted to the performance of a timing circuit often used with the voltage switch.

10.1 THE BJT SWITCH

A BJT voltage switch is usually connected in the CE configuration because this provides power gain and polarity inversion. Both of these features are required to be available in a switching element used in a logic gate. Power gain enables the stage to drive a number of similar stages: polarity inversion enables the output logic levels to be compatible with the input levels and also enables the full range of logical switching functions to be achieved.

Accordingly, the discussion that follows considers the behaviour of a CE switch and, in particular, one using an NPN transistor. Nevertheless, the simple models used to describe it are applicable also to the CB and CE configurations and, with appropriate sign changes, to a PNP device.

The CE switch circuit of Fig. 10.2(a) is a rearrangement of Fig. 10.1(a), with $v_S = V_{CC}$. We now investigate the base-emitter circuit requirements to switch the BJT OFF and ON.

BJT static models

Consider, first, the OFF state of Q. In Chapter 3, the cut-off mode for an NPN transistor was defined by the conditions $V_{CB} > 0$, $V_{BE} \leqslant 0$, as shown by the hatched region of Fig. 10.2(b). However, the region $V_{BE} < 0$ is unnecessarily stringent and is inconvenient in practice. It implies that $V_G < 0$ in Fig. 10.2(a) and this is not possible for a system employing only a positive supply rail. Thus, a relaxed specification for 'cut-off', appropriate to $V_{BE} > 0$, is required.

It is often assumed that the BJT is cut off when $V_{BE} = 0.6$ V but that figure is more appropriate to the 'just-on' state than the 'definitely-off' condition. A figure for that may be deduced from the equation for the transfer characteristic (Equation (3.9), Chapter 3).

Typically, as mentioned in previous bias calculations for the forward-active region, $V_{BE} \sim 0.65$ V for $I_C = 1$ mA for a low power NPN, Si, transistor. Furthermore, a decade change in I_C is accompanied by a change in V_{BE} of 60 mV (approximately). Hence, $I_C = (1 \text{ mA}/1000) = 1 \ \mu\text{A}$ at $V_{BE} = [0.65 \text{ V} - (3 \times 60 \text{ mV})] = 0.47$ V, and $I_C < 0.1 \ \mu\text{A}$ for $V_{BE} = 0.4$ V. This figure

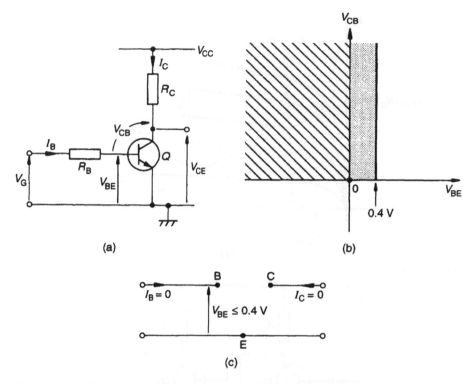

Fig. 10.2 (a) A CE switch. (b) Practical cut-off corresponds to $V_{CB} > 0$, $V_{BE} \leq 0.4$ V. (c) Approximate cut-off model for Q

provides a useful practical *upper limit* for V_{BE} for an acceptable cut-off condition if ambient temperature variations are taken into account. The revised cut-off area in Fig. 10.2(b) now includes the shaded area between $V_{BE} = 0$ and $V_{BE} = 0.4$ V. An approximate d.c. cut-off model for the BJT is shown in Fig. 10.2(c). Consider, next, the ON condition of Q in Fig. 10.2(a). When Q is ON, I_B must be sufficiently large to cause the operating point to move into the saturated region, i.e. to the left of the boundary line $V_{CB} = 0$ (corresponding to $V_{CE} = V_{BE}$) in Fig. 10.3(a), where the potential difference across the BJT is small and denoted by $V_{CE(sat)}$. But there is *no unique value* of $V_{CE(sat)}$: it depends on I_B, for given values of V_{CC} and R_C.

However, in order to be able to interconnect CE switches directly, as in Fig. 10.3(b), so that Q_2 is OFF when Q_1 is ON we must have $V_{CE(sat)} < 0.4$ V. But $V_{CE} = (V_{CB} + V_{BE})$, hence the acceptable operating area in Fig. 10.4 is in the hatched region below the line $V_{CB} = (0.4 \text{ V} - V_{BE})$.

In this saturation region the collector–base junction is forward-biased ($V_{CB} < 0$, or $V_{BC} > 0$) so the simple BJT model that holds for the forward-active region is no longer valid. It is replaced by the first-order model of Fig. 10.5. For this: D_E models the base–emitter junction and I_{BE} is exponentially dependent on V_{BE}; D_C models the collector–base junction and I_{BC} is exponentially dependent on V_{BC}: I_{CT} is exponentially dependent on both V_{BE} and V_{BC} (Getreu, 1976).

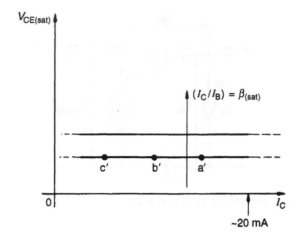

Fig. 10.7   Showing the constancy of $V_{CE(sat)}$ with increase in I_C, for fixed (I_C/I_B)

Figure 10.7 is derived from Fig. 10.6 and is intended to emphasize the constancy of $V_{CE(sat)}$ with increase in I_C when (I_C/I_B) = constant = $\beta_{(sat)}$. Two points deserve mention. First, $V_{CE(sat)}$ is *not* proportional to $\beta_{(sat)}$. Second, the characteristics have, in practice, a small positive slope due to the bulk resistance of the emitter and collector regions of the BJT. This is taken into account in the specification of $V_{CE(sat)}$, which is a maximum for the I_C range considered.

Figure 10.8(a) shows the resulting two-voltage-generator static circuit model for a saturated NPN BJT. The generators are shown as batteries but, of course,

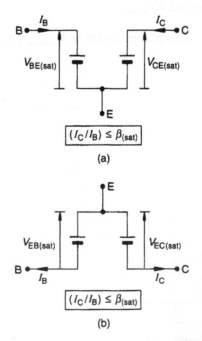

Fig. 10.8   Two-generator model for a saturated transistor, derived from Fig. 10.7. (a) NPN transistor; (b) PNP transistor

they can deliver no power to an external circuit. They merely represent terminal voltage drops when the BJT switch is ON. For completeness, the model of a saturated PNP BJT is shown in Fig. 10.8(b).

Manufacturers do not usually specify $\beta_{(sat)}$ explicitly but they do supply data from which it can be deduced. Thus, the data sheet for the NPN device type BC108 states that, $V_{CE(sat)} < 250$ mV for $I_C = 10$ mA, $I_B = 0.5$ mA and $V_{BE(sat)} = 700$ mV (typical) at these currents. No maximum is given for $V_{BE(sat)}$ but, as we will see later, its value is not as critical as that of $V_{CE(sat)}$.

Allowing a safety margin of 0.1 V and writing $V_{BE(sat)} = 0.8$ V is usually sufficient in design work. From the data given,

$$\beta_{(sat)} = (I_C/I_B) = (10/0.5) = 20, \text{ for } V_{CE(sat)} = 0.25 \text{ V (max)}$$

Hence, we can construct the model shown in Fig. 10.9, which will be used for the remainder of this chapter.

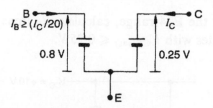

Fig. 10.9   Saturation model for NPN transistor type BC108

Note that $TCV_{BE(sat)} \sim (-2 \text{ mV/}°C)$ because that is the tempco of D_E in Fig. 10.5. Furthermore, $|TCV_{CE(sat)}| \ll 1$ mV/°C because it is the *difference* in the tempcos of D_E and D_C.

Example 10.1 _____

For the BC108 data given above,

(a) calculate the minimum value of I_B to guarantee that $V_{CE(sat)} < 0.25$ V for $I_C = 5$ mA
(b) what happens to $V_{CE(sat)}$ if I_B exceeds the value found in (a)?

Solution

(a) $\beta_{(sat)} = 20$, hence, for $I_C = 5$ mA, I_B (min) = (5 mA/20) = 0.25 mA.
(b) In Fig. 10.10, point Q corresponds to the limit condition in (a).
    If $I_B > 0.25$ mA, the operating points moves to Q'.
    Hence, $V_{CE(sat)}$ *decreases*, but we cannot say by how much. In switching circuit design we are only interested in the *limit* conditions. In digital logic terms, anything below $V_{CE(sat)}$ counts as a '0' in this instance.

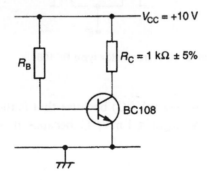

Fig. 10.10   Relating to Example 10.1

Example 10.2

Using a resistor from the E24 range, calculate R_B (max) for the circuit of Fig. 10.11 if the BJT operates with $V_{CE(sat)} \leqslant 0.25$ V.

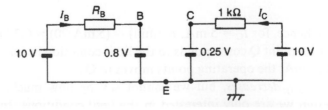

Fig. 10.11   Circuit for Example 10.2

Solution

A simplified equivalent circuit for Fig. 10.11, that incorporates the saturation model of Fig. 10.9, is shown in Fig. 10.12.

Fig. 10.12   Equivalent representation of Fig. 10.11

$I_C(\text{max}) = (10 - 0.25)\text{V}/(1\,\text{k}\Omega - 5\%) = 10.26\,\text{mA}$

$I_B(\text{min}) = I_C(\text{max})/\beta_{(\text{sat})} = 10.26\,\text{mA}/20 = 0.513\,\text{mA}$

But, $I_B(\text{min}) = (10 - 0.8)\,\text{V}/R_B(\text{max})$

$\therefore R_B(\text{max}) = 9.2\,\text{V}/0.513\,\text{mA} = 17.92\,\text{k}\Omega$

Allowing a $\pm 5\%$ tolerance on R_B, the nearest resistor value, in the E24 range, that meets this condition is 16 kΩ. Thus, $R_B(\text{max}) = 16\,\text{k}\Omega \pm 5\%$.

Note that, in saturation, I_C and I_B are both circuit-determined because $V_{CC} \gg V_{BE(\text{sat})}, V_{CE(\text{sat})}$ (*see* Problem 2).

BJT dynamic response

A thorough understanding of BJT dynamic response, and the circuit models and their parameters used to characterize it, requires a detailed knowledge of device physical operating mechanisms. That lies outside the scope of this book, so only a brief qualitative description is given in this section (*See* Sparkes, 1966).

Base charge provides a key to an appreciation of BJT transient response. In a first-order model of switching in the forward-active region, the collector current, i_C, is regarded as instantaneously proportional to the magnitude of the minority carrier charge in transit from the emitter end of the base, where it is injected, to the collector end of the base, where it is extracted. For an NPN transistor in the CE configuration this charge, q_B, is attracted into the base region, across the emitter–base junction by an equal and opposite majority carrier charge, produced in the base by a flow of current in the base lead: i_C can only change as quickly as q_B. Once established, q_B (and, hence, i_C) only remains constant if a steady base current supplies charge lost through the recombination of minority carriers.

We consider Fig. 10.2(a) again, but regard V_G and V_{CE} as time-dependent quantities, v_G, v_{CE}, respectively. Figure 10.13 shows the effect, at the output, for two conditions of a rectangular input waveform, v_G.

For the first condition, shown by the waveforms labelled (i), the amplitude of v_G is insufficient for saturation to occur. The step changes in v_G, at $t = 0$, $t = t_d$, produce corresponding step changes in i_B but q_B is related to the time-integral of i_B so i_C and v_{CE} take time to change. The mechanism at $t = 0$ is analogous to the process of filling a container (the base) with a quantity of water (Q_{B1}) by a constant flow (I_{B1}) of water, from a tap, when the container is leaky (recombination). There is no *sudden* change in water level.

In the second condition of base drive, shown by the waveforms labelled (ii), the amplitude of v_G *is* sufficient for saturation to occur. The waveform for v_{CE} now exhibits a smaller transition time at switch-on but v_{CE} remains at $V_2 = V_{CE(\text{sat})}$ for a time t_s longer than the duration of the base drive waveform.

t_s is the 'storage delay time' and it increases with the amplitude of v_G. An alternative description for t_s is 'excess minority carrier storage time' because the

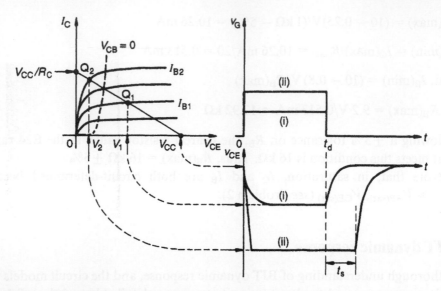

Fig. 10.13 Showing the effect of generator voltage amplitude on the collector voltage waveform, for the circuit of Fig. 10.2(a)

base contains a charge of minority carriers in excess of that required to support the same value of I_C in the unsaturated state.

This excess charge must be removed before I_C can decrease. The penalty paid for a value of $V_{CE(sat)}$ sufficiently small to guarantee cut-off in a CE stage, driven by the stage in question, is thus an output waveform of increased duration. This can limit pulse repetition rates in digital systems so methods have been devised to minimize it.

Two of these are shown in the 'before' and 'after' pictures of Fig. 10.14. Figure 10.14(a) shows an experimental CE saturating switch, of the type under discussion, and observed waveforms: t_s is clearly visible. Figure 10.14(b) shows the effect on v_{CE} of a 'speed-up' capacitor connected across the base resistor. This provides an easy access path for the total base minority carrier charge when the input voltage changes suddenly.

Extending our previous water analogy, the quickest way of supplying water to a container, or of removing water from it, is to use a bucket of appropriate size to dump the water in, or to scoop it out: the bucket and capacitor perform similar functions. There are at least two problems with this scheme: the capacitive loading on the driver stage; and in IC designs a capacitor normally takes up a larger chip area than a diode, transistor or resistor.

Figure 10.14(c) shows how a Schottky diode can be connected to reduce t_s. As indicated in Chapter 1, the threshold-of-conduction voltage for a Schottky diode is about 0.3 V less than that of an ordinary Si PN junction, so connecting the diode as shown reduces the forward bias on the collection–base junction when the BJT saturates. This reduces the excess stored charge and hence t_s the Schottky, being a majority carrier device, exhibits no storage effects. In IC designs the

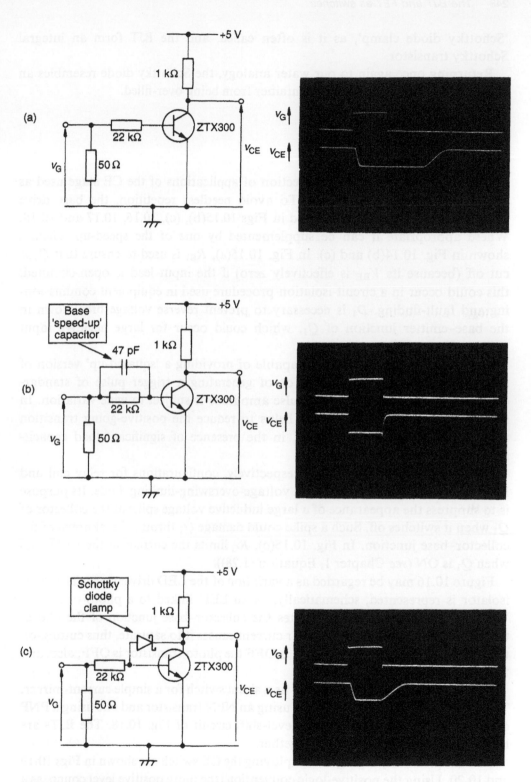

Fig. 10.14   Observed waveforms for v_G, v_{CE} for three types of base drive circuit. Common time and voltage scales apply to all the traces. The upper waveforms in (a), (b) and (c) have a lower level at zero volts, an amplitude of 5 V and a duration of 1 μs

'Schottky diode clamp', as it is often called, and the BJT form an integral Schottky transistor.

Returning once again to our water analogy, the Schottky diode resembles an overflow pipe that prevents the container from being over-filled.

BJT switch applications

Figures 10.15 to 10.20 show a selection of applications of the CE stage used as either a series or shunt switch. To avoid needless repetition, the base drive circuitry of Fig. 10.15(a) is omitted in Figs 10.15(b), (c), 10.16, 10.17 and 10.18. Where appropriate it can be supplemented by one of the speed-up schemes shown in Fig. 10.14(b) and (c). In Fig. 10.15(a), R_{B1} is used to ensure that Q_1 is cut off (because its V_{BE} is effectively zero) if the input lead is open-circuited: this could occur in a circuit-isolation procedure used in equipment commissioning and fault-finding. D_1 is necessary to prevent reverse voltage breakdown in the base–emitter junction of Q_1, which could occur for large negative input voltages.

The circuit of Fig. 10.15(a) is capable of providing a 'squared-up' version of a mains-derived input waveform or of generating a trigger pulse of standardized amplitude when the input pulse amplitude is subject to wide variation. In such cases a 'catching-diode', D_2 helps to reduce the positive-going transition time of the output waveform, v_O, in the presence of significant load capacitance, C_L.

Figure 10.15(b) and (c) show, respectively, configurations for relay coil and LED drive. In Fig. 10.15(b), D_3 is a voltage-overswing-limiting diode. Its purpose is to suppress the appearance of a large inductive voltage spike at the collector of Q_1 when it switches off. Such a spike could damage Q_1 through breakdown of the collector–base junction. In Fig. 10.15(c), R_C limits the current in the LED (D_4) when Q_1 is ON (*see* Chapter 1, Equation (1.28)).

Figure 10.16 may be regarded as a variation of the LED driver stage. The opto-isolator is represented, schematically, as an LED linked to a photo-transistor. When the LED is ON it illuminates the collector–base junction of the photo-transistor and the resulting collector current causes it to saturate, thus cutting-off the following stage. When the LED is OFF the photo-transistor is OFF, also, and the following transistor saturates.

In Fig. 10.17 the CE stage is used as a shunt switch for a simple current-mirror. The interconnection of a CE switch using an NPN transistor and one using a PNP device produces the logic-voltage level-shift circuit of Fig. 10.18. The BJTs are both ON together or both OFF together.

Two simple binary logic gates employing the CE switch are shown in Figs 10.19 and 10.20. Using the positive-logic convention (the more positive level counts as a '1'), Fig. 10.19 shows a two input NOR gate and Fig 10.20 a two-input NAND gate based on diode-transistor logic (DTL).

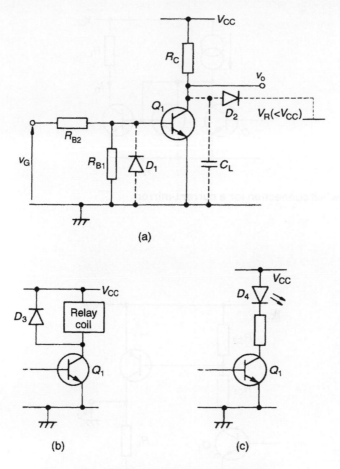

(a)

(b)   (c)

Fig. 10.15   (a) Basic CE switch. (b) Relay coil drive. (c) LED drive

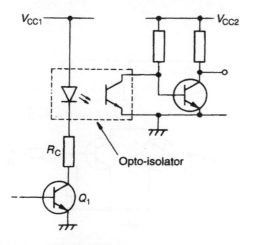

Fig. 10.16   An opto-isolator driven by, and driving, a CE switch

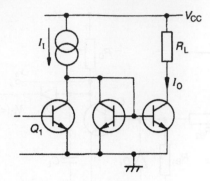

Fig. 10.17   Switch connection for a current-mirror

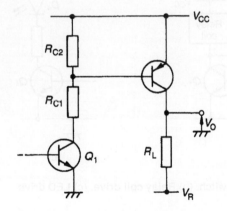

Fig. 10.18   A complementary CE switch used in logic-level shifting

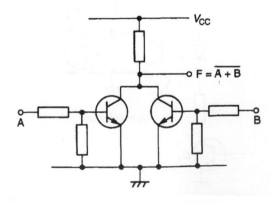

Fig. 10.19   A CE NOR gate

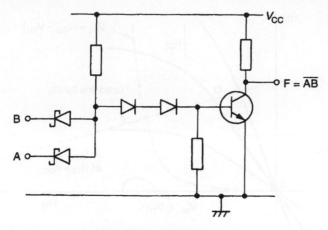

Fig. 10.20 A CE NAND gate (diode-transistor logic)

10.2 THE FET SWITCH

The JFET and MOSFET are both used as switches. However, the enhancement-mode MOSFET is generally used in preference to the JFET because it is capable of exhibiting a lower ON resistance, is easier to drive and simpler to produce in IC form. Consequently, we will concentrate on the MOSFET switch and, in particular, the N-channel device but the discussion is relevant also to a P-channel device if appropriate changes are made in the polarities of device parameters and the designation of terminal currents and voltages.

The MOSFET switch: static characteristics

Consider the N-channel common-source (CS) switching circuit of Fig. 10.21(a). For $V_{GS} < V_{TH}$, where V_{TH} is the threshold-voltage defined in Chapter 5, the MOSFET is cut off.

$$\therefore I_{DS} = I_{LN}. \tag{10.2}$$

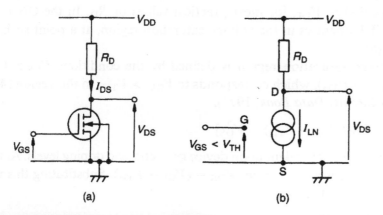

(a)   (b)

Fig. 10.21 (a) An N-channel MOSFET switch. (b) Cut-off condition for (a)

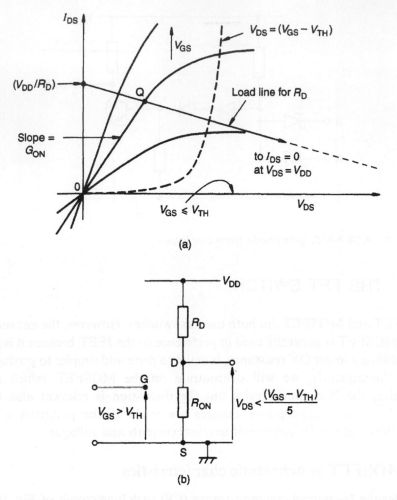

Fig. 10.22 (a) Magnified view of voltage-saturation region. (b) Circuit representation of (a)

I_{LN} is a small ($<1\,\mu A$), temperature-sensitive leakage current, the subscript L referring to leakage and N to N-channel.

The equivalent OFF circuit is shown in Fig. 10.21(b): $V_{DS} = (V_{DD} - I_{LN}R_D) \simeq V_{DD}$ for most practical values of R_D. In the ON condition the MOSFET operates in the voltage-saturation region, at a point such as Q in Fig. 10.22(a).

The voltage saturation region is defined by the conditions $V_{GS} > V_{TH}$ and $V_{DS} < (V_{GS} - V_{TH})$, which corresponds to $V_{GD} > V_{TH}$. In this region (*McMOS Integrated Circuits Data Book*, 1973),

$$I_{DS} = K_M[2V_{DS}(V_{GS} - V_{TH}) - V_{DS}^2] \qquad (10.3)$$

K_M is a parameter dependent upon device geometry and doping levels. At the edge of the voltage-saturation region, $V_{DS} = (V_{GS} - V_{TH})$. Substituting this value of V_{DS} into Equation (10.3) gives,

$$I_{DS} = K_M(V_{GS} - V_{TH})^2 \qquad (10.4)$$

This equation holds for the current saturation region (if the Early effect is ignored) and has the same form as Equation (5.2). Suppose, in Equation (10.3), that,

$$V_{DS}^2 \ll 2V_{DS}(V_{GS} - V_{TH}) \tag{10.5}$$

Then,

$$I_{DS} \simeq 2K_M V_{DS}(V_{GS} - V_{TH}) \tag{10.6}$$

The ON conductance, G_{ON}, and resistance, R_{ON}, are given by,

$$G_{ON} = 1/R_{ON} = (I_{DS}/V_{DS}) = 2K_M(V_{GS} - V_{TH}) \tag{10.7}$$

The condition expressed in (10.5) is valid in practice if,

$$V_{DS}^2 < [2V_{DS}(V_{GS} - V_{TH})/10] \tag{10.8}$$

i.e.,

$$V_{DS} < (V_{GS} - V_{TH})/5 \tag{10.9}$$

Taking the order-of-magnitude figures $V_{GS} = 5$ V, $V_{TH} = 2.5$ V, Equation (10.7) is valid for $V_{DS} < 0.5$ V.

In the vicinity of the origin the MOSFET, when ON, behaves as a (gate) voltage-controllable ohmic resistor and the equivalent circuit of Fig. 10.22(b) is applicable to the CS switch.

Suppose, by choice of R_D, that

$$(V_{DD}/I_{LN}) \gg R_D \gg R_{ON} \tag{10.10}$$

Then, the MOSFET can be regarded as a short-circuit when ON and an open-circuit when OFF. This is the case with the logic circuits dealt with in a later section. The magnitude of G_{ON} and its variation with gate-source voltage, V_{GS}, (*see* Fig. 10.23) is important in the design of voltage-tunable filters and voltage-variable attenuators. However, for analogue switches the variation of G_{ON} with V_{GS} is undesirable. Figure 10.24 shows a series-connected N-channel MOSFET

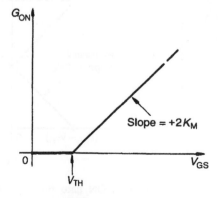

Fig. 10.23   Conductance plot for Fig. 10.22

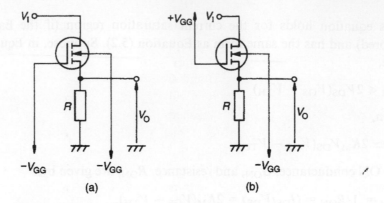

(a)   (b)

Fig. 10.24 A series analogue switch showing gate voltage connection for: (a) the OFF state, (b) the ON state

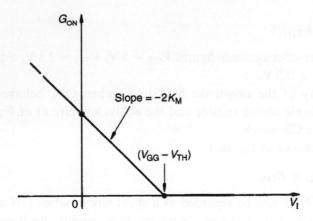

Fig. 10.25 Conductance plot for Fig. 10.24(b)

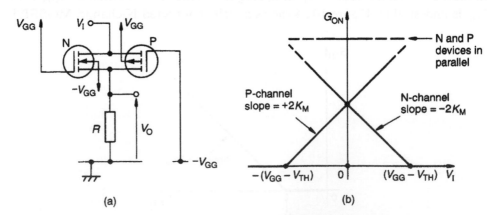

(a)   (b)

Fig. 10.26 (a) A 'transmission' gate with ON gate bias. (b) Conductance plots for a transmission gate analogue switch, the input being V_I and the output V_O. In (a), $V_G = -V_{GG}$ the MOSFET is OFF and V_O is effectively zero: in (b), $V_G = +V_{GG}$ the MOSFET is ON and $V_{DS}(=V_I - V_O)$ is only a few millivolts. Hence, $V_O \simeq V_I$, $V_{GS} = (V_{GG} - V_O) \simeq (V_{GG} - V_I)$ and $G_{ON} = 2K_M[(V_{GG} - V_{TH}) - V_I]$. A plot of this is shown in Fig. 10.25. It is valid for $V_I < 0$ provided the conditions $V_{GD} > V_{TH}$, $V_{GS} > V_{TH}$ are still satisfied.

The problem now arises that G_{ON} varies over the range of the input signal and this leads to distortion in the output. Thus, if $V_{GG} = 10$ V, $V_{TH} = 5$ V and $V_I \simeq 0$ V then $G_{ON} \simeq 10K_M$, but if $V_I = 2$ V then $G_{ON} = 6K_M$.

A way of minimizing this effect is to use a specially constructed 'transmission gate' (Fig. 10.26(a)), in which a P-channel device is connected in parallel with the N-channel MOSFET. The operation of this can be understood by assuming that the P-channel MOSFET has characteristics which are, respectively, ideal complements of those of the N-channel MOSFET. In that case, a plot of its G_{ON} vs V_I characteristic is a mirror-image of that of the N-channel device, as shown in Fig. 10.26(b). Consequently, the net parallel conductance of the pair remains constant over the signal range. This is a simplification because ideal complementary devices are not available and because the threshold voltage of each device is dependent, to some extent, on gate-substrate voltage. Nevertheless, there is a significant reduction in the variation of G_{ON} with V_I for the parallel-connected MOSFET configuration.

In the symbol for the transmission gate, shown in Fig. 10.27, the interlocked oppositely-directed arrowheads are indicative of the bilateral nature of the device (i.e. its ability to transmit signals in each direction). The terminal marked C (Control) is connected to the gate of the constituent N-channel MOSFET. When C = '1' the transmission gate is ON and when C = '0' it is OFF.

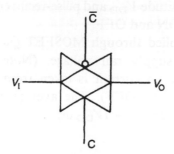

Fig. 10.27 Circuit symbol for a transmission gate: C is connected to the gate of the N-channel MOSFET

The MOSFET switch: dynamic considerations

Whether a MOSFET switch is OFF or ON, the drain-source circuit is shunted by the inter-electrode capacitance, C_{ds}, in addition to any load capacitance connected between these terminals.

The input capacitance is dependent ON the gate–source capacitance, C_{gs}, and the gate–drain capacitance, C_{gd}, which is subject to magnification by the Miller Effect (Chapter 7) when the switch is in the transition mode between the OFF and ON states (or, vice-versa).

The effect of these capacitances on the operation of particular circuits is dealt with, where applicable, in the next section.

MOSFET switch applications

Devices of a single polarity, either N-channel (NMOS) or P-channel (PMOS) are employed for large scale digital storage. PMOS is easier to make but NMOS is faster. Both types of device are used in Complementary MOS logic (CMOS). Some of the application areas in which NMOS (or PMOS) logic is advantageous in comparison to CMOS logic, and vice-versa, are shown by the dot entries in Table 10.1. Table 10.2 illustrates the operation of the basic inverter, which is at the heart of CMOS logic. When A = '0', Q_N is OFF, Q_P is ON and the potential difference ($I_{LN}R_{ON}$) across it is only of the order of 1 mV. Hence the '1' output corresponds to V_{DD} (approximately). Similarly, when A = '1', Q_P is OFF and Q_N is ON. In this case the '0' output corresponds to zero volts (approximately) for $V_{SS} = 0$ V. The input current under d.c. conditions is only of the order of pA, so there is virtually no change in output voltage levels when the inverter is loaded by a large number of similar stages.

The static power dissipation, P_S, is given by $I_{LP}V_{DD}$. This is, typically, less than 5 μW and completely masked by the dynamic dissipation, P_D.

P_D is dependent upon the effective input capacitance, C_i (~ 5 pF), of a stage. In Fig. 10.28(a) the inverter drives n similar stages so $C_L = nC_i$. Figure 10.28(b) shows the current, i, that flows from the supply rail, V_{DD}, when a rectangular input waveform, v_G, of amplitude V_{DD} and pulse-recurrence frequency $f (= I/T)$ is used to switch the inverter ON and OFF.

A charge $C_L V_{DD}$ is supplied through MOSFET Q_P in each switching cycle to charge C_L up to the supply rail voltage. (Note that C_L is *discharged* through MOSFET Q_N but no current is taken from V_{DD} when this occurs because MOSFET Q_P is then OFF). The average current, $\bar{i}$, per cycle is $(C_L V_{DD}/T) = C_L V_{DD}f$. Hence, $P_D = \bar{i}V_{DD}$ or,

$$P_D = fC_L V_{DD}^2 \qquad (10.11)$$

Table 10.1 Comparison table for MOSFET logic: a dot entry denotes a comparative advantage

	Manufacture	IC packing density	Cost/ function	Power supplies	Dissipation	Speed	Typical applications
NMOS or PMOS	•	•	•				Large scale storage
CMOS				•	•	•	General logic

Table 10.2 The CMOS inverter and its operation

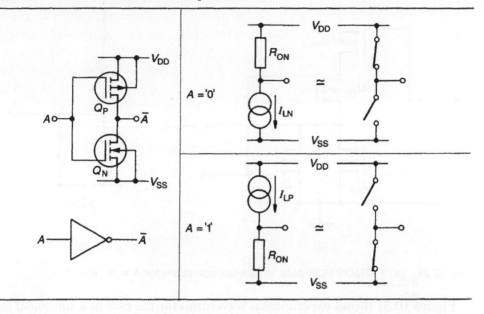

To implement an *n*-input NOR gate in CMOS, *n* N-channel MOSFETs are connected with their sources and drains, respectively, in parallel. The corresponding P-channel devices are connected in series and compromise the drain load of the composite N-channel structure.

Figure 10.29(a) shows a two-input NOR gate and Fig. 10.29(b) its equivalent switch representation for A = '0', B = '1'. In such gates, the substrates of the N-channel devices are taken to V_{SS} (earth potential, in this case) and the substrates of the P-channel devices are connected to V_{DD}.

To implement a NAND gate, the P-channel devices are connected in parallel and the N-channel devices in series.

In Fig. 10.30, the time-selection pulse for a transmission gate is supplied by a saturating complementary switch (Q_1, Q_2, and associated resistors), of a type previously shown (Fig. 10.18), which may be driven by TTL.

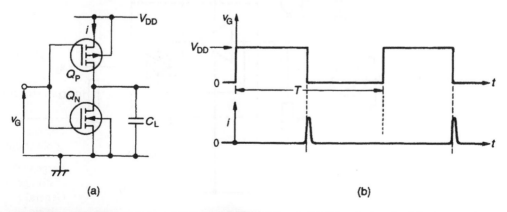

Fig. 10.28 (a) CMOS inverter driving a capacitive load. (b) Waveforms for (a)

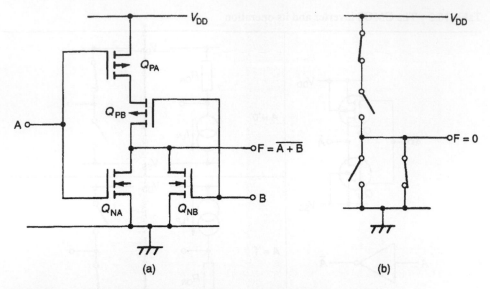

Fig. 10.29   (a) A CMOS NOR gate. (b) Switch conditions for A = '0', B = '1'

Figure 10.31 shows experimental waveforms for the case of a sinusoidal input waveform, v_I.

Figure 10.32 shows the basic elements of a four-input analogue multiplexer. The input signals v_1, v_2, v_3, v_4 are sampled sequentially by data selection pulses applied to the individual gating inputs (C, in Fig. 10.27) of the switches in a 'quad' package. The 'always closed' switch, in series with the feedback resistor in the OA signal-mixing stage, reduces the gain error by balancing the R_{ON} of the input switches.

In Fig. 10.33, (a) shows a simple sample-and-hold circuit and (b) illustrates its mode of operation.

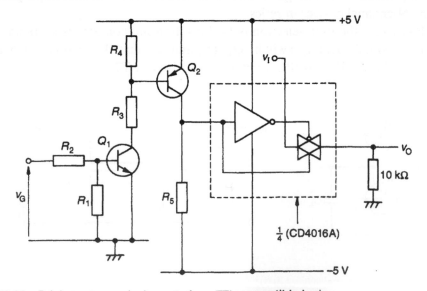

Fig. 10.30   Driving a transmission gate from TTL-compatible logic

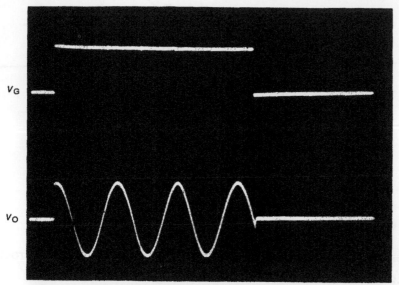

Fig. 10.31   Waveforms for Fig. 10.30. Upper trace v_G, lower trace v_O. The traces have common time and voltage scales. The amplitude of the gating pulse (v_G) is 2 V and its duration is approximately 100 μs

At $t = t'$, a short duration sampling pulse v_C, switches on the transmission gate and the storage capacitor, C_S, charges up to the level of the signal voltage, v_G, via the R_{ON} of the switch. At the end of the sampling interval the gate switches off and the capacitor charge can only decay slowly as a result of the small leakage current of the switch and the input current of the (FET) OA voltage-follower which is included to supply an output voltage signal.

10.3   SWITCH TIMING

The time taken for a capacitor to charge, or discharge, through a specified voltage range via a resistor is of fundamental importance in the analysis and design of circuits using BJT and FET switches and is the subject of this section.

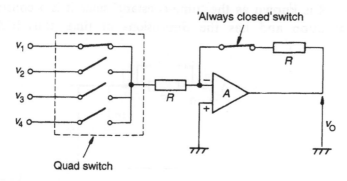

Fig. 10.32   Analogue multiplexer using transmission gates

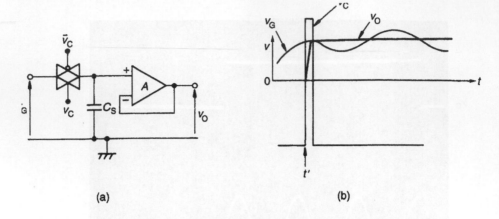

Fig. 10.33 (a) Sample-and-hold circuit. (b) Circuit waveforms for (a)

Consider the circuit of Fig. 10.34. Initially the potential difference across C is zero but V_G is switched into circuit at $t = 0$. Then,

$$V_G = v_R + v_C \qquad (10.12)$$

But,

$$v_C = q/C \qquad (10.13)$$

and,

$$i = (dq/dt) \qquad (10.14)$$

$$\therefore v_R = iR = R(dq/dt) \qquad (10.15)$$

From Equations (10.12), (10.13) and (10.15),

$$V_G = R(dq/dt) + (q/C) \qquad (10.16)$$

The solution of this, subject to the condition $q = 0$ at $t = 0$ is,

$$q = CV_G[1 - e^{-t/CR}] \qquad (10.17)$$

The product CR is known as the 'time-constant' since it is a constant for the circuit configuration and has the dimensions of time: thus $[CR] = [(Q/V)$

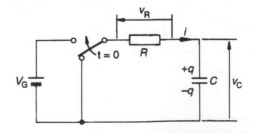

Fig. 10.34 Circuit for switching analysis

$(V/I)] = [Q/I] = [t]$. Physically, CR is an indicator of the time required to change levels.

It follows from Equations (10.13) and (10.17), that,

$$v_C = (q/C) = V_G[1 - e^{-t/CR}] \qquad (10.18)$$

Substituting this expression for v_C in Equation (10.12) yields,

$$v_R = V_G - V_G[1 - e^{-t/CR}] = V_G e^{-t/CR} \qquad (10.19)$$

Furthermore,

$$i = (v_R/R) = (V_G/R) e^{-t/CR} \qquad (10.20)$$

Figure 10.35(a) shows the charging curve for v_C: that for q is identical in shape, differing only by a vertical scale factor C. Figure 10.35(b) shows the waveform for v_R, which is complementary to that of v_C (i.e. $v_R + v_C = \text{constant} = V_G$). The curve for i is identical in shape to it but differs by a vertical scale factor R.

Providing we make the definitions illustrated in Fig. 10.35, it is possible to describe v_C, v_R, by the *same* formula, namely,

remainder = span × exponential decay factor (10.21)

'Span' is defined as the magnitude of the difference between the 'target' (or 'aiming') level and the 'start' (or 'initial') level.

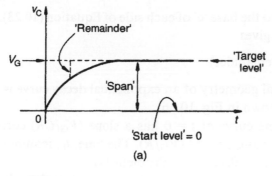

(a)

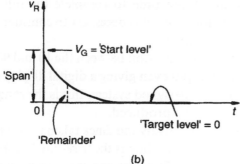

(b)

Fig. 10.35 Waveforms for Fig. 10.34: (a) v_C; (b) v_R

The 'start' level is that which exists immediately after a switching action has occurred. In determining this we note that the potential difference across a capacitor cannot change instantaneously. Thus, if the potential difference between the plates is 'x' volts immediately prior to a switching action, then it is also 'x' volts immediately after the switching action.

The 'target' level is that which would be reached after an infinite time. It is a d.c. level and can be calculated by imagining the capacitor to be removed from the circuit.

The 'remainder' is the magnitude of the difference between the target level and the level actually reached at a particular time.

The 'exponential decay factor' is $e^{-t/CR}$. In this the *minus* sign denotes *decay*: a plus sign denotes growth. The appropriate value of CR, if not obvious by visual inspection of the circuit configuration, can be obtained by the application of Thévenin's Theorem (*see* Example 10.4 below).

Problems in timing circuits are usually of two types. In the first, the problem is to calculate the level reached after a given time: Equation (10.21) is then directly applicable. In the second, the problem is to calculate the time to reach a specified level. In this case it is necessary to transpose Equation (10.21) to obtain a formula that is directly usable. Thus,

$$1/(\text{exponential decay factor}) = \text{span/remainder} \qquad (10.22)$$

or,

$$e^{+t/CR} = \text{span/remainder} \qquad (10.23)$$

Taking logarithms to the base 'e' of each side of Equation (10.23), and multiplying throughout by CR, gives

$$t = CR \log_e (\text{span/remainder}) \qquad (10.24)$$

Some of the essential geometry of an exponential decay curve is illustrated in the waveform for v_C, shown in Fig. 10.36.

The tangent to the curve at $t = 0$ has a slope (V_G/CR) corresponding to an initial capacitor charging current (V_G/R). The time, t_h, required to complete half the span is given by $t_h \simeq 0.7CR$ (*see* Problem 12).

Theoretically, it takes an infinite time to complete the full span. Hence, in defining a meaningful 'transition time', it is necessary to consider the time taken to cover a specified fraction of the span.

It is conventional to specify that fraction between the 10 and 90% levels. This is easy to measure (some oscilloscopes even giving a digital display of the times) and is practical too since, in any well-designed system, this is the range within which a subsequent switching action will be produced.

In Fig. 10.36, t_1 and t_2, are respectively the times taken to complete 10 and 90% of the span. In this case the transition time is the 'rise-time', t_r. For the waveform of v_R, t_1 is the 90% point, t_2 the 10% point and the transition time is called the 'fall-time', $t_f(=t_r)$. For practical purposes the span is regarded as having been

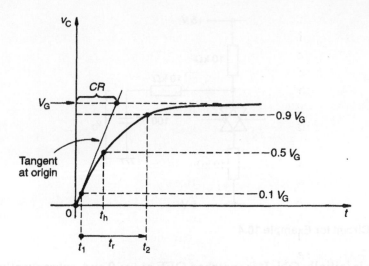

Fig. 10.36 Features of the exponential decay curve

fully traversed after a time equal to five time-constants ($5CR$) has elapsed, since the remainder is then less than 1% of the span.

Example 10.3

Derive an expression for t_r in Fig. 10.36.

Solution

The span is V_G. At $t = t_1$, remainder $= (V_G - 0.1V_G) = 0.9V_G$
Using Equation (10.24),

$$t_1 = CR \log_e (V_G/0.9V_G) = CR \log_e (1/0.9)$$

At $t = t_2$, remainder $= (V_G - 0.9V_G) = 0.1V_G$

$$\therefore t_2 = CR \log_e (V_G/0.1V_G) = CR \log_e (1/0.1)$$

Hence,

$$t_r = (t_2 - t_1) = [CR \log_e (1/0.1)] - [CR \log_e (1/0.9)]$$

Using the identity ($\log_e a - \log_e b$) $\equiv \log_e (a/b)$,

$$t_r = CR \log_e 9 \simeq 2.2CR.$$

Example 10.4

In the circuit of Fig. 10.37 the transmission gate can be taken as having zero resistance when ON and zero leakage when OFF.

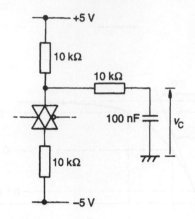

Fig. 10.37   Circuit for Example 10.4

The gate is initially ON. It is switched OFF at $t = 0$ and, subsequently switched ON 20 ms later.

(a) Sketch and dimension the waveform v_C.
(b) Calculate v_C at $t = 7.5$ ms.
(c) Calculate the times at which $v_C = -2.5$ V.

Solution

(a) In order to sketch v_C it is necessary, first, to draw equivalent circuits for the ON and OFF conditions of the gate.

In Fig. 10.38, (a) shows the full circuit when the gate is ON and (b) shows a simplified form of it, obtained by the use of Thévenin's Theorem. Under d.c. conditions the current in these resistors is zero and initially $v_C = -5$ V.

Figure 10.39 shows the equivalent circuit when the gate is OFF. The target voltage for v_C is $+5$ V. When the gate is switched OFF the charging time constant is, by inspection of Fig. 10.39, (100 nF × 20 kΩ) = 2 ms: when the gate is subsequently switched ON the discharging time-constant, obtained from Fig. 10.38(b), is (100 nF × 15 kΩ) = 1.5 ms.

Figure 10.40 is a sketch of waveform v_C. P_1 corresponds to $t = t_1 = 7.5$ ms and P_2, P_3 to $v_C = -2.5$ V.

(b) The span is 10 V. Using Equation (10.21), remainder = $10\,e^{-7.5/2} = 0.235$ V. Since the target level is $+5$ V, the level reached at P_1 is $(5 - 0.235)$V = 4.765 V. As 20 ms > $5CR$, the target level *is* reached during the switching pulse.

(c) At P_2, remainder = $5 - (-2.5)$ V = 7.5 V
Using Equation (10.24), the time at P_2 is,

$$t_2 = 2 \times \log_e (10/7.5) \text{ ms}$$

or,

$$t_2 = 0.575 \text{ ms}$$

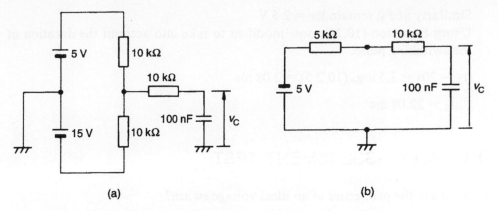

(a)   (b)

Fig. 10.38   (a) ON-state equivalent circuit for Fig. 10.37. (b) Simplified form of (a)

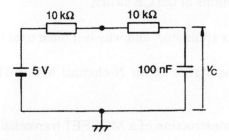

Fig. 10.39   OFF-state equivalent circuit for Fig. 10.37

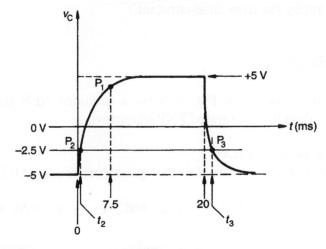

Fig. 10.40   Capacitor voltage waveform for Fig. 10.37

Similarly at P_3, remainder = 2.5 V
Using Equation (10.24), now modified to take into account the duration of the switching pulse,

$(t_3 - 20) = 1.5 \log_e (10/2.5) = 2.08$ ms

$\therefore t_3 = 22.08$ ms

10.4 SELF ASSESSMENT TEST

1 What are the properties of an ideal voltage switch?

2 Is there a unique value for the collector–emitter saturation voltage of a BJT?

3 What is the definition of the parameter $\beta_{(sat)}$?

4 State three applications of the CE switch.

5 What is a 'Schottky transistor' and why is it often used in CE switches?

6 How is the ON resistance of an N-channel MOSFET dependent on gate voltage?

7 What is the basic construction of a MOSFET transmission gate?

8 What is the function of an analogue signal multiplexer?

9 If the supply rail voltage of a capacitively-loaded CMOS inverter is doubled, by what factor does the switching power dissipation change?

10 What is meant by the term 'time-constant'?

10.5 PROBLEMS

(Use the parameter values in Fig. 10.9 for a saturated NPN transistor and corresponding values for a saturated PNP transistor).

1   A CE switch has $I_C = 5$ mA.
    Calculate the minimum value of I_B that guarantees $V_{CE(sat)} < 0.25$ V

2   Show that if, in Fig. 10.11, $V_{BE(sat)}$ and $V_{CE(sat)}$ are both negligible in comparison with V_{CC}, then,

$R_B \leqslant \beta_{(sat)} R_0$

3 Figure 10.41 shows part of a proximity-detector. When a surface bearing a magnet comes close to the reed-relay (RR), the relay contacts close, Q saturates and acts as a sink for the load current, I_L.

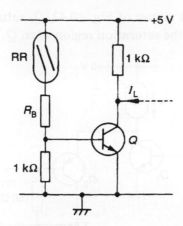

Figure 10.41

Calculate a suitable value for R_B, when $I_L = 5$ mA if the current in the relay contacts is not to exceed 5 mA.

4 Figure 10.42 shows the circuit of a rudimentary design for a water-level alarm system. When the water level in a container rises sufficiently high for the probes P_1 and P_2 to become immersed, Q_P saturates and supplies a load current I_L to an alarm indicator denoted by R_C.

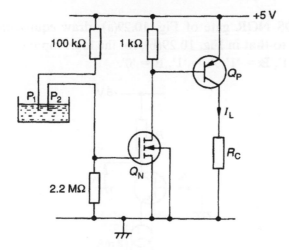

Figure 10.42

(a) Show that, when ON, Q_N always operates in the current-saturation region, for which $I_{DS} = K_M(V_{GS} - V_{TH})^2$.

(b) Calculate the maximum permitted value of I_L
(Assume: $K_M = 200\ \mu A/V^2$; $V_{TH} = 2\ V$; zero resistance between the probes when they are immersed).

5   In the logic-level-shift circuit of Fig. 10.43, Q_1 saturates when ON but Q_2 is not intended to enter the saturation region when Q_1 is OFF.

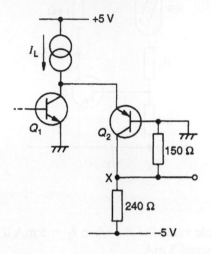

Figure 10.43

(a) Calculate the maximum value of I_L that satisfies this condition.
(b) Find the voltage level at point X when Q_2 is ON and $I_2 = 10\ mA$.
(Assume that $\alpha = 0.99$ for Q_2)

6   Calculate I_B and V_C for the circuit of Fig. 10.44 for $V_{CE(sat)} = 0.25\ V$. Hence, calculate $R_C(min)$ for this condition.

7   For the CMOS NOR gate of Fig. 10.29(a), draw equivalent switch circuit states, similar to that in Fig. 10.29(b), for the logic input conditions: A = '0', B = '0'; A = '1', B = '1'; A = '1', B = '0'.

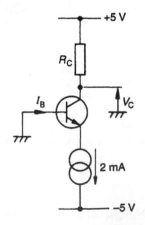

Figure 10.44

8 Draw a circuit diagram of a two-input CMOS NAND gate.

9 In Fig. 10.45 a square voltage waveform of amplitude V_G is applied between the gate and source terminals of MOSFET Q_N to chop, for subsequent a.c. amplification, the output voltage of a temperature transducer, shown inside the dotted contour.

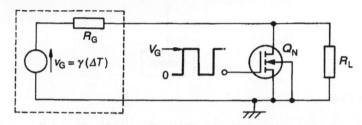

Figure 10.45

When ON, Q_N is characterized by Equation (10.7) of the text: when it is OFF, $I_{LN} = 0$.

Calculate, from the following data, the amplitude of the square wave developed across the load R_L: $\gamma = 40\ \mu V/°C$; $\Delta T = 50°C$; $K_M = 100\ \mu A/V^2$; $V_G = 8\ V$; $V_{TH} = 3\ V$; $R_G = 1\ k\Omega$; $R_L = 1\ M\Omega$.

10 Show on a diagram how four MOSFET transmission gates can be interconnected to implement the double-pole double-throw (DPDT) switch function of Fig. 10.46.

11 Show, by differentiation, that Equation (10.17) leads to Equation (10.16).

12 Referring to Fig. 10.36.

   (a) Show that $t_h \simeq 0.7CR$.
   (b) Calculate v_C for $t = CR$.
   (c) Calculate t for $v_C = 2V_G/3$.

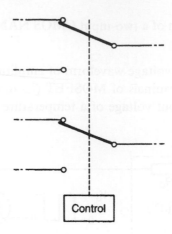

Figure 10.46

13 In a practical version of Fig. 10.33(a), $C_S = 1000$ pF and $R_{ON} = 200\ \Omega$ for the transmission gate.

Calculate the minimum duration, t_d, of the gate sampling pulse for v_O to attain a value within 1% of a fixed input voltage, V_G.

(Assume the OA is ideal in all respects).

14 Establish the validity of the expressions for v_O in column 3 of Table 7.2 (Chapter 7).

(Hint: in row (a) replace V, R_g, R_i, by a Thévenin equivalent circuit).

15 Use the series expansion of e^{-x} to show that, for $t \ll CR$,

$$V_G\, e^{-t/CR} \simeq V_G[1 - (t/CR)]$$

16 In Fig. 10.47 the transmission gate, which may be assumed to have the characteristics of an ideal voltage switch, is switched off at $t = 0$. Calculate the time at which $i_C = 0.5$ mA.

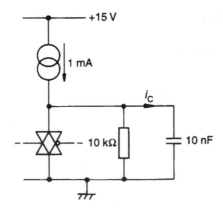

Figure 10.47

Answers to numerical problems

Chapter 1
1 (a) $I = 2.65\ \mu A$; (b) $V_D = 0.75\ V$; (c) $|\Delta V_D| = 57.6\ mV$. **2** $I_1 = 0.73\ mA$.
3 $V_{DI} = -17.3\ mV$. **4** $\theta_X = (1/b)\log_e 2$; $\theta_X = 9.9°C$. **9** $C_1 \geqslant 0.2\ \mu F$;
$C_2 \geqslant 0.02\ \mu F$; $I = 0.1\ mA$. **10** $I_L(max) = 61\ mA$; $R_S = 150\ \Omega$.
13 $\Delta V_o = \pm 1.2\ mV$ (assume the shunt regulator output section does not load the
input section). **14** $f_R(max)/f_R(min) = 3$. **15** $R_S = 270\ \Omega$.

Chapter 2
1 $R = 216\ k\Omega$. **4** 10%. **7** 17.5 V; > 17.5 V; 2.14 A.

Chapter 3
4 $P_C(max) = 250\ mW$. **5** $I_C = 1.89\ mA$; $V_C = 7.61\ V$; $I_C(max) = 1.93\ mA$;
$I_C(min) = 1.87\ mA$. **6** $I_{CI} = 2.15\ mA$. **7** $R_E = (51\ k\Omega \pm 1\%) + (4.7 k\Omega \pm 1\%)$;
$R_V = 10\ k\Omega$ (linear); $R_X(max) = 80.5\ k\Omega$. **8** (a) 1.81 mA, 10.11 V; (b) 1.74 mA,
10.3 V; (c) 1.24 mA, 11.65 V; Discrepancies arise through different values of
R_{TH}. **9** $I_C = 2.52\ mA$, $V_C = 8.2\ V$ for the following choice: $R_E = 620\ \Omega$;
$R_1 = 8.2\ k\Omega$; $R_2 = 43\ k\Omega$; $R_C = 2.7\ k\Omega$. **11** $I_1 \geqslant 10 I_B$. **12** $I_C = 1.03\ mA$ at
$V_{XX} = 5\ V$; $I_C = 1.08\ mA$ at $V_{XX} = 10\ V$.

Chapter 4
4 $r_\pi = 500\ \Omega$; $g_m = 200\ mS$; $r_o = 20\ k\Omega$; $r_e = 5\ \Omega$; $\alpha = 0.99$; $r_c = 2\ M\Omega$.
6 $(V_o/V_s) = -270$; $R_o = 4.3\ k\Omega$. **7** $(V_o/V_s) = -204$. **8** $(V_o/V_s) = +37.7$; 23.4 Ω.

Chapter 5
1 $|I_G| = 181\ nA$. **2** $V_{GS} = +2.5\ V$. **3** $K_J = 160\ \mu A/V^2$. **5** $V_P = -5.2\ V$.
6 $I_{DSS} = 6\ mA$, $V_P = -6\ V$. **8** $I_{DSQ} = 1\ mA$ (approx.). **9** $(V_{Z1} - V_{Z2}) =$
7.3 mV; $V_{DD} = 14\ V$. **10** $I_{DSQ}(max) = 1.17\ mA$; $I_{DSQ}(min) = 1.00\ mA$.
12 $I_{DSQ}(max) = 0.62\ mA$; $I_{DSQ}(min) = 0.41\ mA$. **13** One suitable design is:
$R_1 = 47\ k\Omega \pm 1\%$; $R_2 = 100\ k\Omega \pm 1\%$; $R_3 = 10\ M\Omega \pm 5\%$; $R_S = 11\ k\Omega \pm 1\%$.

Chapter 6
5 $g_{fs} = 1.09\ mS$; $r_{ds} = 110\ k\Omega$; $\mu = 120$. **6** $r_{ds} = 52.5\ k\Omega$; $g_{fs} = 2.9\ mS$.
7 $A_v = -16.5$. **9** $(V_o/V_g) = 11.4$. **14** R_X and R_V do not have unique values:
suitable values are; $R_X = 100\ k\Omega$, $R_V = 1\ k\Omega$ (linear). Reading error is 1%
(approx.).

Chapter 7

2 (a) $|A| = 100/[1 + 4f^2]$; $\angle A = -2 \tan^{-1} 2f$. (b) $\angle A = -90°$.
6 (a) $A = -50/[1 - j(0.1/f)][1 + j(0.01f)]$, where f is in kHz. (b) At $f = 25$ Hz,
$|A| = 12.13$, $\angle A = 256°$; at $f = 300$ kHz, $|A| = 15.81$, $\angle A = 108.43°$. **8** $|G_{m.f.}|_{dB} =$
25.18; $\omega_{s1} = 20$ rad/s; $\omega_{s2} = 60$ rad/s; $\omega_g = 10$ rad/s; $\omega_H = 23 \times 10^6$ rad/s.
9 $f_T = 500$ MHz; $f_\beta = 5$ MHz. **10** $r_\pi = 1$ kΩ; $C_\pi = 29.83$ pF; $r_o = 40$ kΩ.
11 $f_H = 11.6$ MHz (approx.). **12** $f_L = 318$ Hz; $f_H = 17.5$ MHz.

Chapter 8

1 $A = 1050$; $\beta = -0.019$. **5** $G = -25/[1 - j(0.01/f)][1 + j(f/2000)]$, where f is in
Hz. **6** $R_i = 1$ MΩ; $R_o = 10$ Ω; $f_H' = 1$ MHz. **10** $G = 47.14 \angle -45°$.
11 $(f/f_H) = \sqrt{3}$; $|A\beta| = A_o\beta_o/8$.

Chapter 9

1 For the circuit of Fig. 9.2(a), $R_1 = 10$ kΩ $\pm 1\%$, $R_2 = 100$ kΩ $\pm 1\%$.
2 $v_o = av_1 + bv_2 - cv_3 - dv_4$. For $a = 1$, $b = 5$, $c = 2$, $d = 3$, the choice
$R = 75$ kΩ gives $(R/a) = 75$ kΩ and $(R/b) = 15$ kΩ. The choice $R_1 = 36$ kΩ gives
$(R_1/c) = 18$ kΩ, $(R_1/d) = 12$ kΩ. **3** $R_1 = 100$ kΩ $\pm 1\%$; $R_2 = 1$ MΩ $\pm 1\%$;
$R_3 = 10$ kΩ $\pm 1\%$; $R_4 = (180$ kΩ $\pm 1\%)\|(180$ kΩ $\pm 1\%)$.
4 $R = (220$ kΩ $\pm 1\%) + (270$ kΩ $\pm 1\%)$; $r = 10$ kΩ $\pm 1\%$. **6** Referring to Fig.
9.12 (a): $R_1 = 10$ kΩ $\pm 1\%$; $R_2 = (39$ kΩ $\pm 1\%) + (1$ kΩ $\pm 1\%)$. Connect two
1 MΩ $\pm 5\%$ resistors in parallel between amplifier input terminal and chassis-
earth. **7** $R_1 = (68$ kΩ $\pm 1\%) + (12$ kΩ $\pm 1\%)$; $R_2 = 10$ kΩ $\pm 1\%$;
$R_3 = (68$ kΩ $\pm 1\%) + (22$ kΩ $\pm 1\%)$. **8** Input resistance is 20 kΩ/V.
10 $R_1 = R_3 = 10$ kΩ $\pm 1\%$; $R_2 = R_4 = 100$ kΩ $\pm 1\%$. **11** $R_X = 16$ kΩ.
12 $I_o = 10$ nA. **13** $V_o = -\alpha V/2$. **14** $v_O = (a + 1)(v_A - v_B)$.
16 $R \simeq 18$ kΩ. **17** $R_1 = 16$ kΩ $\pm 1\%$; $C_1 = 0.5$ μF; $R_2 = 160$ kΩ $\pm 1\%$;
$C_2 = 50$ pF. **18** $V_{OS} = 1.22$ mV (assume: 10 kΩ + 10 Ω = 10 kΩ;
$I_{OS} \times 10$ $\Omega \ll 1.22$ mV). **19** Referring to Equation (9.43), the term
$(-I_N R \pm V_{OS})$ is replaced by $\pm(I_{OS}R + V_{OS})$. **20** $G = 4.14 \angle 114.4°$.
21 $f \simeq 8$ kHz.

Chapter 10

1 $I_B(\text{min}) = 0.25$ mA. **3** $(3$ kΩ $\pm 5\%) > R_B > (910$ Ω $\pm 5\%)$.
4 (a) When Q_N is ON, $V_{DS} = 4.2$ V, $V_{GS} = 4.78$ V, hence $V_{DS} > (V_{GS} - V_{TH})$;
(b) $I_L(\text{max}) = 14.97$ mA. **5** (a) $I_L(\text{max}) = 21$ mA; (b) -1 V (approx.).
6 $I_B = 95.2$ μA; $V_C = -0.55$ V; $R_C(\text{min}) = 2.91$ kΩ. **9** 1 mV (approx.).
12 (b) $0.63 V_G$; (c) $t \simeq 1.1CR$. **13** $t_d \simeq 5CR = 1$ μs. **16** $t = 69.3$ μs.

References and further reading

References

Getreu I. 1976: *Modeling the bipolar transistor.* Beaverton, Oregon: Tektronix Inc.

Gray PR & Meyer RG. 1977: *Analysis and design of analog integrated circuits.* Chichester: John Wiley & Sons, 50–4.

Hayt WH & Kemmerly JE. 1993: *Engineering circuit analysis* (5th edition). London: McGraw-Hill.

Motorola Inc. 1973: *McMOS integrated circuits data book.*

Muller RS & Kamins TI. 1977: *Device electronics for integrated circuits.* Chichester: John Wiley & Sons, 200–18.

Powell R. 1995: *Introduction to electric circuits.* London: Edward Arnold.

Sparkes JJ. 1966: *Junction transistors.* Oxford: Pergamon Press.

Till WC & Luxon JT. 1982: *Integrated circuits: Materals, devices and fabrication.* London: Prentice Hall, 112–36.

Further reading

The comments in brackets are those of the author of the present volume.

Millman J & Grabel A. 1987: *Microelectronics* (2nd edition). London: McGraw-Hill Book Company. (An undergraduate textbook of encyclopaedic scope which pursues at greater depth concepts, e.g. feedback, introduced in this volume.)

Horowitz P & Hill W. 1989: *The art of electronics* (2nd edition). Cambridge: Cambridge University Press. (This reference book has a bias towards circuit design and contains a wealth of information on device and component data and tried-and-tested circuits.)

Safety handbook for undergraduate electrical teaching laboratories. Produced by the Department of Electrical Engineering, University of Southampton (England) in association with the Health and Safety Executive. HSE Books, PO Box 1999, Sudbury, Suffolk, CO10 6FS, England. (An essential safety booklet for all users of electrical/electronic laboratories.)

Appendix

RESISTOR DATA

To provide adequate variation, while keeping the number of stock types at a reasonable level, the electronics industry long ago standardized on a number of *preferred* values in each decade range of (discrete component) resistance value.

Preferred values are based on a geometric progression that minimizes overlaps and gaps in the range when tolerances are taken into account. The ratio between adjacent values is approximately 10 to the power $(1/n)$, where n is the number in the range. The relationship between n and tolerance $t(\%)$ is, $t = (\pm 120/n)$.

Table A1 shows values for the popular E6, E12, E24 ranges which have, respectively, 6, 12, 24 values in each decade. Although the E24 range refers to a $\pm 5\%$ tolerance, component manufacturers also supply values in this range with a $\pm 1\%$ tolerance. The E48 and higher ranges are not widely used because of stock problems.

Table A2 shows the standard resistor colour-band-coding system and Table A3 an example of its use.

An alternative to the standard number-letter code is the BS1852 code. In this the Ω sign is dispensed with and R, K, M take the place of the decimal point in the approprite decade range.

Tolerance is indicated also by letters: F($\pm 1\%$); G($\pm 2\%$); J($\pm 5\%$); K($\pm 10\%$); M($\pm 20\%$). A suggested mnemonic for remembering this is, *Few Good Judges Know Much.*

Examples of complete resistor descriptions are given in Table A4.

Table A1 Popular 'E' ranges for resistors

E6	$\pm 20\%$ range	10		15		22		33		47		68		
E12	$\pm 10\%$ range	10	12	15	18	22	27	33	39	47	56	68	82
E24	$\pm\ 5\%$ range	10 11	12 13	15 16	18 20	22 24	27 30	33 36	39 43	47 51	56 62	68 75	82 91

Table A2 The resistor colour-coding scheme

Note: band order reads from end to centre of resistor

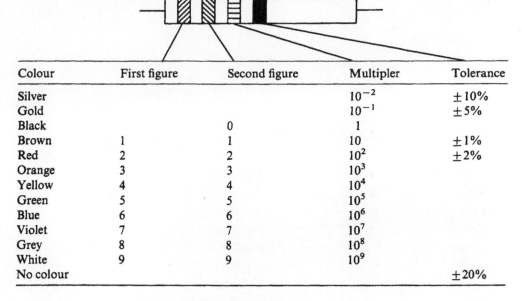

Colour	First figure	Second figure	Multipler	Tolerance
Silver			10^{-2}	$\pm10\%$
Gold			10^{-1}	$\pm5\%$
Black		0	1	
Brown	1	1	10	$\pm1\%$
Red	2	2	10^2	$\pm2\%$
Orange	3	3	10^3	
Yellow	4	4	10^4	
Green	5	5	10^5	
Blue	6	6	10^6	
Violet	7	7	10^7	
Grey	8	8	10^8	
White	9	9	10^9	
No colour				$\pm20\%$

Table A3 Colour-code examples

Example ($10\ k\Omega \pm 10\%$)

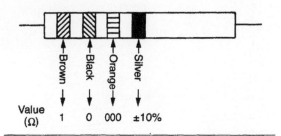

Other examples:

Red, violet, orange, silver	$27\,000 \pm 10\%$
Blue, black, brown, gold	$600\ \Omega \pm 5\%$
Brown, black, gold	$1\ \Omega \pm 20\%$
Orange, orange, silver	$0.33\ \Omega \pm 20\%$

Table A4 Examples of the BS 1852 code

Older code	BS 1852
$0.22\ \Omega \pm 20\%$	R22M
$1\ \Omega \pm 10\%$	1R0K
$2.7\ \Omega \pm 5\%$	2R7J
$1\ k\Omega \pm 1\%$	1K0F
$8.2\ k\Omega \pm 2\%$	8K2G
$56\ k\Omega \pm 10\%$	56KK
$6.8\ M\Omega \pm 20\%$	6M8M

Index

Printed and bound by CPI Group (UK) Ltd, Croydon, CR0 4YY

03/10/2024

01040001 9999